Discrete-Time Inverse Optimal Control for Nonlinear Systems

System of Systems Engineering Series

Series Editor: Mo Jamshidi

Discrete-Time Inverse Optimal Control for Nonlinear Systems

Edgar N. Sanchez • Fernando Ornelas-Tellez

CRC Press
Taylor & Francis Group
Boca Raton London New York

CRC Press is an imprint of the
Taylor & Francis Group, an **Informa** business

Part of cover image from G. Quiroz, Bio-Signaling Weighting Functions to Handle Glycaemia in Type 1 Diabetes Mellitus via Feedback Control System (in Spanish). PhD thesis, IPICYT (Potosinian Institute of Scientific and Technological Research), San Luis Potosi, Mexico, 2008. With permission.

CRC Press
Taylor & Francis Group
6000 Broken Sound Parkway NW, Suite 300
Boca Raton, FL 33487-2742

First issued in paperback 2017

© 2013 by Taylor & Francis Group, LLC
CRC Press is an imprint of Taylor & Francis Group, an Informa business

No claim to original U.S. Government works

Version Date: 20130222

ISBN 13: 978-1-4665-8087-9 (hbk)
ISBN 13: 978-1-138-07381-4 (pbk)

Library of Congress Cataloging-in-Publication Data

Sanchez, Edgar N.
    Discrete-time inverse optimal control for nonlinear systems / authors, Edgar N. Sanchez, Fernando Ornelas-Tellez.
        pages cm
    Includes bibliographical references and index.
    ISBN 978-1-4665-8087-9 (hardback)
    1. Nonlinear control theory. 2. Neural networks (Computer science) 3. Discrete-time systems. 4. Kalman filtering. I. Ornelas-Tellez, Fernando. II. Title.

QA402.35.S36 2013
003'.83--dc23   2012050923

Visit the Taylor & Francis Web site at
http://www.taylorandfrancis.com

and the CRC Press Web site at
http://www.crcpress.com

Dedication

To my wife María de Lourdes and

our sons, Zulia Mayari, Ana María and Edgar Camilo

Edgar N. Sanchez

To my wife Yenhi and my son Alexander

Fernando Ornelas-Tellez

Contents

List of Figures

List of Tables

Preface

Optimal nonlinear control is related to determining a control law for a given system, such that a cost functional (performance index) is minimized; it is usually formulated as a function of the state and input variables. The major drawback for optimal nonlinear control is the need to solve the associated Hamilton–Jacobi–Bellman (HJB) equation. The HJB equation, as far as we are aware, has not been solved for general nonlinear systems. It has only been solved for the linear regulator problem, for which it is particularly well-suited.

This book presents a novel inverse optimal control for stabilization and trajectory tracking of discrete-time nonlinear systems, avoiding the need to solve the associated HJB equation, and minimizing a cost functional. Two approaches are presented; the first one is based on passivity theory and the second one is based on a control Lyapunov function (CLF). It is worth mentioning that if a continuous-time control scheme is real-time implemented, there is no guarantee that it preserves its properties, such as stability margins and adequate performance. Even worse, it is known that continuous-time schemes could become unstable after sampling.

There are two advantages to working in a discrete-time framework: a) appropriate technology can be used to implement digital controllers rather than analog ones; b) the synthesized controller is directly implemented in a digital processor. Therefore, the control methodology developed for discrete-time nonlinear systems can be implemented in real systems more effectively. In this book, it is considered a class of nonlinear systems, the affine nonlinear systems, which represents a great variety of systems, most of which are approximate discretizations of continuous-time systems.

The main characteristic of the inverse optimal control is that the cost functional is determined a posteriori, once the stabilizing feedback control law is established. Important results on inverse optimal control have been proposed for continuous-time linear and nonlinear systems, and the discrete-time inverse optimal control has been analyzed in the frequency domain for linear systems. Different works have illustrated adequate performances of the inverse optimal control due to the fact that this control scheme benefits from adequate stability margins, while the minimization of a cost functional ensures that control effort is not wasted.

On the other hand, for realistic situations, a control scheme based on a plant model cannot perform as desired, due to internal and external disturbances, uncertain parameters and/or unmodelled dynamics. This fact motivates the development of a model based on recurrent high order neural networks (RHONN) in order to identify the dynamics of the plant to be controlled. A RHONN model is easy to implement, has relatively simple structure and has the capacity to adjust its parameters on-line. This book establishes a neural inverse optimal controller combining two techniques: a) inverse optimal control, and b) an on-line neural identifier, which uses a recurrent neural network, trained with an extended Kalman filter, in order to determine a model for an assumed unknown nonlinear system.

The applicability of the proposed controllers is illustrated, via simulations, by stabilization and trajectory tracking for nonlinear systems. As a special case, the proposed control scheme is applied to glycemic control of type 1 diabetes mellitus (T1DM) patients in order to calculate the adequate insulin delivery rate in order to prevent hyperglycemia and hypoglycemia levels.

This book is organized as follows.

- Chapter 1 introduces the discrete-time inverse optimal control for nonlinear systems is given.

- Chapter 2 briefly describes useful results on optimal control theory, Lyapunov stability and passivity theory, required in future chapters in order to provide the inverse optimal control solution. Additionally, this chapter presents a neural scheme in order to identify uncertain nonlinear systems.

- Chapter 3 deals with inverse optimal control via passivity for both stabilization and trajectory tracking of discrete-time nonlinear systems. The respective inverse optimal controller is synthesized. Also, the inverse optimal control for a class of positive nonlinear systems is established. Examples illustrate the proposed control scheme applicability.

- Chapters 4 and 5 establish the inverse optimal control and its solution by using a quadratic control Lyapunov function, in order to ensure the stabilization of discrete-time nonlinear systems. These results are extended to achieve trajectory tracking along a desired reference. Additionally, an inverse optimal trajectory tracking for block-control form nonlinear systems and positive systems is proposed. Simulation results illustrate the applicability of the proposed control schemes.

- Chapter 6 discusses the combination of the results presented in Chapter 2, Chapter 3 and Chapter 5 in order to achieve stabilization and trajectory tracking for uncertain nonlinear systems, by using a RHONN scheme to model uncertain nonlinear systems, and then applying the inverse optimal control methodology. Examples illustrate the applicability of the proposed control techniques.

- Chapter 7 presents a case study: glycemic control of type 1 diabetes mellitus patients.

- Chapter 8 gives concluding remarks.

Edgar N. Sanchez

Fernando Ornelas-Tellez

Guadalajara and Morelia, Mexico

Acknowledgments

The authors thank CONACyT (for its name in Spanish, which stands for National Council for Science and Technology), Mexico, for financial support on projects 131678 and 174803 (Retencion). They also thank CINVESTAV-IPN (for its name in Spanish, which stands for Advanced Studies and Research Center of the National Polytechnic Institute), Mexico, particularly Rene Asomoza Palacio, President of the CINVESTAV system, and Bernardino Castillo-Toledo, Director Guadalajara Campus, and Universidad Michoacana de San Nicolas de Hidalgo, Morelia, Mexico, in particular J. Jesus Rico-Melgoza, J. Aurelio Medina-Rios and Elisa Espinosa-Juarez, for encouragement and facilities provided to accomplish this book publication. In addition, they thank Mo Jamshidi, University of Texas at San Antonio, USA, Alexander G. Loukianov, CINVESTAV, Guadalajara Campus, Alma Y. Alanis and Eduardo Ruiz-Velazquez, Universidad de Guadalajara, Mexico, Pedro Esquivel-Prado, Instituto Tecnologico de Tepic, Mexico, Ph.D. students Blanca S. Leon and Jose Santiago Elvira-Ceja, CINVESTAV, Guadalajara Campus, and M.Sc. students Guillermo C. Zuñiga and Gabriel Casarrubias, Universidad Michoacana de San Nicolas de Hidalgo, who have contributed to this book achievement. Last but not least, they appreciate the understanding and patience of their families during this book writing.

Authors

Edgar N. Sanchez was born in 1949, in Sardinata, Colombia, South America. He obtained the BSEE, major in Power Systems, from Universidad Industrial de Santander (UIS), Bucaramanga, Colombia in 1971, the MSEE from CINVESTAV-IPN (Advanced Studies and Research Center of the National Polytechnic Institute), major in Automatic Control, Mexico City, Mexico, in 1974 and the Docteur Ingenieur degree in Automatic Control from Institut Nationale Polytechnique de Grenoble, France in 1980. Since January 1997, he has been with CINVESTAV-IPN, Guadalajara Campus, Mexico. He was granted a USA National Research Council Award as a research associate at NASA Langley Research Center, Hampton, Virginia, USA (January 1985 to March 1987). He is also a member of the Mexican National Research System (promoted to highest rank, III, in 2005), the Mexican Academy of Science and the Mexican Academy of Engineering. He has published more than 100 technical papers in international journals and conferences, and has served as reviewer for different international journals and conferences. He has also been a member of many international conferences IPCs, both IEEE and IFAC. His research interest centers on neural networks and fuzzy logic as applied to automatic control systems.

Fernando Ornelas-Tellez was born in Patzcuaro, Michoacan, Mexico, in 1981. He received the BSEE degree from Morelia Institute of Technology in 2005, and the M.Sc. and D.Sc. degrees in Electrical Engineering from the Advanced Studies and

Research Center of the National Polytechnic Institute (CINVESTAV-IPN), Guadala-

jara Campus, in 2008 and 2011, respectively. Since 2012 he has been with Michoacan

University of Saint Nicholas of Hidalgo, where he is currently a professor of Electrical

Engineering graduate programs. His research interest centers on neural control, direct

and inverse optimal control, passivity and their applications to biomedical systems,

electrical machines, power electronics and robotics.

Notations and Acronyms

NOTATIONS

$\forall$	for all
$\in$	belonging to
$\Rightarrow$	implies
$\subset$	contained in
$\subseteq$	contained in or equal to
$\cup$	union
$\cap$	intersection
$:=$	equal by definition
$\lambda_{min}(Q)$	the minimum eigenvalue of matrix Q
$\lambda_{max}(Q)$	the maximum eigenvalue of matrix Q
$P > 0$	a positive definite matrix P
$P \geq 0$	a positive semidefinite matrix P
ΔV	Lyapunov difference
$\geq$	larger than or equal to
$\leq$	less than or equal to
$\succeq$	larger than or equal to, component wise
$\preceq$	less than or equal to, component wise

$\mathscr{A}$	set or vector space
$\mathscr{K}$	a class $\mathscr{K}$ function
$\mathscr{K}_\infty$	a class $\mathscr{K}_\infty$ function
$\mathscr{K}\mathscr{L}$	a class $\mathscr{K}\mathscr{L}$ function
$\mathbb{N}$	the set of all natural numbers
$\mathbb{Z}^+$	the set of nonnegative integers
$\mathbb{R}$	the set of all real numbers
$\mathbb{R}^+$	the set of positive real numbers
$\mathbb{R}_{\geq 0}$	the set of nonnegative real numbers
$\mathbb{R}^n$	n-dimension vector space
$(\cdot)^T$	matrix transpose
$(\cdot)^{-1}$	matrix inverse
$(\cdot)^*$	optimal function
C^ℓ	ℓ-times continuously differentiable function
$\|x\|_n$	the n-norm of vector x
$|x|$	absolute value of vector x
$\|x\|$	the Euclidean norm of vector x
$\alpha_1 \circ \alpha_2$	the composition of two functions, where $\alpha_1(\cdot) \circ \alpha_2(\cdot) = \alpha_1(\alpha_2(\cdot))$

ACRONYMS

ANN	Artificial Neural Network
AR	Autoregressive
ARX	Autoregressive External
BIBS	Bounded-Input Bounded-State
CLF	Control Lyapunov Function
CT	Continuous-Time
DARE	Discrete-Time Algebraic Riccati Equation
DC	Direct Current
DM	Diabetes Mellitus
DOF	Degrees of Freedom
DT	Discrete-Time
EKF	Extended Kalman Filter
FPIOC	Inverse Optimal Control with Fixed Parameters
GAS	Globally Asymptotically Stable
GS	Globally Stable
GSC	Grid Side Converter
HJB	Hamilton-Jacobi-Bellman
HJI	Hamilton-Jacobi-Isaacs
ISS	Input-to-State Stable
LQR	Linear Quadratic Regulator
MI	Matrix Inequality

MPC	Model Predictive Control
MPILC	Model Predictive Iteration Leaning Control
NN	Neural Network
NNARX	Neural Network Auto regressive External
PBC	Passivity-Based Control
PID	Proportional Integral Derivative
PWM	Pulse-Width Modulation
RHONN	Recurrent High Order Neural Network
RHS	Right-Hand Side
RMLP	Recurrent Multilayer Perceptron
RNN	Recurrent Neural Network
RSC	Rotor Side Converter
SG	Speed-Gradient
SG-IOC	Speed-Gradient Inverse Optimal Control
SG-IONC	Speed-Gradient Inverse Optimal Neural Control
SGUUB	Semiglobally Uniformly Ultimately Bounded
T1DM	Type 1 Diabetes Mellitus

1 Introduction

This chapter presents fundamental issues on the optimal control theory. The Hamilton–Jacobi–Bellman (HJB) equation is introduced as a means to obtain the optimal control solution; however, solving the HJB equation is a very difficult task for general nonlinear systems. Then, the inverse optimal control approach is proposed as an appropriate alternative methodology to solve the optimal control, avoiding the HJB equation solution.

Optimal control is related to finding a control law for a given system such that a performance criterion is minimized. This criterion is usually formulated as a cost functional, which is a function of state and control variables. The optimal control can be solved using Pontryagin's maximum principle (a necessary condition) [105], and the method of dynamic programming developed by Bellman [12, 13], which leads to a nonlinear partial differential equation named the HJB equation (a sufficient condition); nevertheless, solving the HJB equation is not an easy task [56, 121]. Actually, the HJB equation has so far rarely proved useful except for the linear regulator problem, to which it seems particularly well suited [8, 67].

This book presents a novel inverse optimal control for stabilization and trajectory tracking of discrete-time nonlinear systems, which are affine on the control input, avoiding the need to solve the associated HJB equation, and minimizing a cost func-

1

tion. For the inverse optimal control in the continuous-time setting, we refer the reader to the results presented in [8, 34, 36, 46, 77, 85, 135]. The discrete-time inverse optimal control has been treated in the frequency domain for linear systems in [8, 136], based on the return difference function. Although there already exist many important results on inverse optimal control for continuous-time nonlinear systems, the discrete-time case is seldom analyzed [1, 93], in spite of its advantages for real-time implementation. Additionally, it is not guaranteed that a continuous-time control scheme preserves its properties when implemented in real time; even worse, it is known that continuous-time schemes could become unstable after sampling [89].

For the inverse approach, a stabilizing feedback control law, based on a priori knowledge of a control Lyapunov function (CLF), is designed first, and then it is established that this control law optimizes a cost functional. The main characteristic of the inverse approach is that the cost functional is a posteriori determined for the stabilizing feedback control law.

The existence of a CLF implies stabilizability [56] and every CLF can be considered as a cost functional [21, 36, 52]. The CLF approach for control synthesis has been applied successfully to systems for which a CLF can be established, such as feedback linearizable, strict feedback and feed-forward ones [37, 108]. However, systematic techniques for determining CLFs do not exist for general nonlinear systems.

This book presents two approaches for solving inverse optimal control; the first one is based on passivity theory and the second one is synthesized using a control Lyapunov function.

Additionally, in this book a *robust* inverse optimal control scheme is proposed in order to avoid the solution of the Hamilton–Jacobi–Isaacs (HJI) equation associated with the optimal control for disturbed nonlinear systems. Furthermore, a neural inverse optimal controller based on passivity theory and neural networks is established in order to achieve stabilization and trajectory tracking for uncertain discrete-time nonlinear systems.

In the following, basic fundamentals concerning inverse optimal control are presented.

1.1 INVERSE OPTIMAL CONTROL VIA PASSIVITY

The concepts of passivity and dissipativity for control systems have received considerable attention lately. These concepts were introduced initially by Popov in the early 1950s [106] and formalized by Willems in the early 1970s [134] with the introduction of the storage and supply rate functions. Dissipative systems present highly desirable properties, which may simplify controller analysis and synthesis. Passivity-based control (PBC) was introduced in [97] to define a controller synthesis methodology, which achieves stabilization by passivation.

Passivity is an alternative approach for stability analysis of feedback systems [49]. One of the passivity advantages is to synthesize stable and robust feedback controllers. Despite the fact that nonlinear passivity for continuous-time has attracted considerable attention, and many results in this direction have been obtained [44, 96, 97, 98, 129, 134, 138] and references therein, there are few for discrete-time nonlinear systems [18, 59, 83, 87].

For a continuous-time framework, the connection between optimality and passivity was established by Moylan [84] by demonstrating that, as in the linear case, the optimal system has infinite gain margins due to its passivity property with respect to the output. The passivity property for nonlinear systems can be interpreted as a phase property [121, 96], analogous to linear systems, which is introduced to guard against the effects of unmodeled dynamics (fast dynamics) which cause phase delays [121].

In this book, we avoid solving the associated HJB equation by proposing a novel inverse optimal controller for discrete-time nonlinear systems based on a quadratic storage function, which is selected as a discrete-time CLF candidate in order to achieve stabilization by means of passivation through output feedback under detectability conditions. Moreover, a cost functional is minimized. The CLF acts as a Lyapunov function for the closed-loop system. Finally, the CLF is modified in order to achieve asymptotic tracking for given reference trajectories.

1.2 INVERSE OPTIMAL CONTROL VIA CLF

Due to the fact that the optimal control solution by Bellman's method is associated with solving an HJB equation, the inverse optimal control via a CLF approach is proposed in this book. For this approach, the control law is obtained as a result of solving the Bellman equation. Then, a CLF candidate for the obtained control law is proposed such that it stabilizes the system; a posteriori a cost functional is minimized.

For this book, a quadratic CLF candidate is used to synthesize the inverse optimal control law. Initially, the CLF candidate depends on a fixed parameter to be selected in order to obtain the solution for inverse optimal control. A posteriori, this parameter is adjusted by means of the speed-gradient (SG) algorithm [33] in order to establish the stabilizing control law and to minimize a cost functional. We refer to this combined approach as the *SG inverse optimal control*. The use of the SG algorithm within the control loop is a novel contribution of this approach. Although the SG has been successfully applied to control synthesis for continuous-time systems, there are very few results of the SG algorithm application to stabilization purposes for nonlinear discrete-time setting [88].

On the other hand, considering that systems are usually uncertain in their parameters, exposed to disturbances, and that there exist modeling errors, it is desirable to obtain a robust optimal control scheme. Nevertheless, when we deal with the robust optimal control, in which a disturbance term is involved in the system, the HJI partial differential solution is required. A control law as a result of the robust optimal control formulation and the associated HJI solution provides stability, optimality and robustness with respect to disturbances [36]; however, determining a solution for the HJI equation is the main drawback of the robust optimal control; this solution may not exist or may be extremely difficult to solve in practice.

To overcome the need for the HJI solution, in this book a robust inverse optimal control approach for a class of discrete-time disturbed nonlinear systems is proposed, which does not require solving the HJI equation and guarantees robust stability in the presence of disturbances; additionally, a cost functional is minimized.

1.3   NEURAL INVERSE OPTIMAL CONTROL

For realistic situations, a controller based on a plant model could not perform as desired, due to internal and external disturbances, uncertain parameters, or unmodeled dynamics [38]. This fact motivates the need to derive a model based on recurrent high order neural networks (RHONN) to identify the dynamics of the plant to be controlled. Also, a RHONN model is easy to implement, of relatively simple structure, robust in the presence of disturbances and parameter changes, has the capacity to adjust its parameters on-line [113, 119], and allows incorporating a priori information about the system structure [116]. Three recent books [107, 113, 116] have reviewed the application of recurrent neural networks for nonlinear system neural identification and control. In [116], an adaptive neural identification and control scheme by means of on-line learning is analyzed, where stability of the closed-loop system is established based on the Lyapunov function method. For this neural scheme, an assumed uncertain discrete-time nonlinear system is identified by a RHONN model, which is used to synthesize the inverse optimal controller. The neural learning is performed on-line through an extended Kalman filter (EKF) as proposed in [116].

Finally, as a special case of this book, attention is drawn to identification and control of the blood glucose level for Type 1 diabetes mellitus (T1DM) patients. T1DM is a metabolic disease caused by destruction of the pancreas insulin-producing cells. Diabetes mellitus is one of the costliest health problems in the world and one of the major causes of death worldwide. A nonlinear identification scheme based on neural networks is used as an alternative to determine mathematical models, then the obtained neural model is used for control purposes.

1.4 MOTIVATION

Optimal control laws benefit from adequate stability margins, and the fact that they minimize a cost functional ensures that control effort is not wasted [27, 35]. Indeed, optimal control theory is introduced in [121] as a synthesis tool which guarantees stability margins. On the other hand, robustness achieved as a result of the optimality is largely independent of the selected cost functional [121]. Stability margins characterize basic robustness properties that well designed feedback systems must possess.

In this book, motivated by the favorable stability margins of optimal control systems, we propose an inverse optimal controller, which achieves stabilization and trajectory tracking for discrete-time nonlinear systems, and avoids the HJB equation solution. On the other hand, the inverse optimal control methodology can be applied to uncertain nonlinear systems, modeled by means of a neural identifier, and therefore a robust inverse optimal control scheme is obtained.

For this book, we consider *a class* of discrete-time nonlinear systems, which are affine in the input. Models of this type describe a great variety of systems. Most of them represent approximate discretizations of continuous-time systems [88]. There are two reasons to employ the discrete-time (DT) framework. First, appropriate technology can be used to implement digital controllers rather than analog ones, which are generally more complicated and expensive. Second, the synthesized controller is directly implemented in a digital processor.

To appreciate the importance of this book contribution, one should recall that other methods for selecting the control law, based on the cancellation or domination of non-

linear terms (such as feedback linearization, block control, backstepping technique,

and other nonlinear feedback designs), do not necessarily possess desirable properties

of optimality and may lead to poor robustness and wasted control effort [35]. Even

worse, certain nonlinear terms can represent nonlinear positive feedback, which can

have catastrophic effects in the presence of modeling or measurement errors [34].

Other approaches such as variable structure technique may lead to wasted control

effort.

2 Mathematical Preliminaries

This chapter briefly describes useful results on optimal control theory, Lyapunov stability, passivity and neural identification, required in future chapters, for the inverse optimal control solution. Section 2.1 gives a review of optimal control, prior to the introduction of the inverse optimal control. Section 2.2 presents general stability analysis, and robust stability results are included in Section 2.3 for disturbed nonlinear systems. Section 2.4 establishes concepts related to passivity theory. Finally, Section 2.5 presents a neural scheme in order to identify uncertain nonlinear systems.

2.1 OPTIMAL CONTROL

This section briefly discusses the optimal control methodology and its limitations.

Consider the affine-in-the-input discrete-time nonlinear system:

$$x_{k+1} = f(x_k) + g(x_k)u_k, \qquad x_0 = x(0) \tag{2.1}$$

where $x_k \in \mathbb{R}^n$ is the state of the system at time $k \in \mathbb{Z}^+ \cup 0 = \{0, 1, 2, \ldots\}$, $u_k \in \mathbb{R}^m$ is the input, $f : \mathbb{R}^n \to \mathbb{R}^n$ and $g : \mathbb{R}^n \to \mathbb{R}^{n \times m}$ are smooth mappings, $f(0) = 0$ and $g(x_k) \neq 0$ for all $x_k \neq 0$.

For system (2.1), it is desired to determine a control law $u_k = \bar{u}(x_k)$ which minimizes the following cost functional:

$$V(x_k) = \sum_{n=k}^{\infty} \left(l(x_n) + u_n^T R u_n \right) \tag{2.2}$$

where $V : \mathbb{R}^n \rightarrow \mathbb{R}^+$ is a performance measure [51]; $l : \mathbb{R}^n \rightarrow \mathbb{R}^+$ is a positive semidefinite[1] function weighting the performance of the state vector x_k, and $R : \mathbb{R}^n \rightarrow \mathbb{R}^{m \times m}$ is a real symmetric and positive definite[2] matrix weighting the control effort expenditure. The entries of R could be functions of the system state in order to vary the weighting on control efforts according to the state value [51].

Equation (2.2) can be rewritten as

$$
\begin{aligned}
V(x_k) &= l(x_k) + u_k^T R u_k + \sum_{n=k+1}^{\infty} l(x_n) + u_n^T R u_n \\
&= l(x_k) + u_k^T R u_k + V(x_{k+1}).
\end{aligned} \tag{2.3}
$$

From Bellman's optimality principle [11, 67], it is known that, for the infinite horizon optimization case, the value function $V^*(x_k)$ becomes time invariant and satisfies the discrete-time Bellman equation [2, 11, 91]

$$
V^*(x_k) = \min_{u_k} \left\{ l(x_k) + u_k^T R u_k + V^*(x_{k+1}) \right\}. \tag{2.4}
$$

Note that the Bellman equation is solved backwards in time [2].

In order to establish the conditions that the optimal control law must satisfy, we define the discrete-time Hamiltonian $\mathscr{H}(x_k, u_k)$ ([41], pages 830–832) as

$$
\mathscr{H}(x_k, u_k) = l(x_k) + u_k^T R u_k + V^*(x_{k+1}) - V^*(x_k) \tag{2.5}
$$

which is used to obtain control law u_k by calculating

$$
\min_{u_k} \mathscr{H}(x_k, u_k).
$$

---

[1] A function $l(z)$ is a positive semidefinite (or nonnegative definite) function if for all vectors z, $l(z) \geq 0$. In other words, there are vectors z for which $l(z) = 0$, and for all others z, $l(z) > 0$ [51].

[2] A real symmetric matrix R is positive definite if $z^T R z > 0$ for all $z \neq 0$ [51].

The value of u_k which achieves this minimization is a feedback control law denoted as $u_k = \bar{u}(x_k)$, then

$$\min_{u_k} \mathcal{H}(x_k, u_k) = \mathcal{H}(x_k, \bar{u}(x_k)).$$

A necessary condition, which this feedback optimal control law $\bar{u}(x_k)$ must satisfy [51], is

$$\mathcal{H}(x_k, \bar{u}(x_k)) = 0. \tag{2.6}$$

$\bar{u}(x_k)$ is obtained by calculating the gradient of the right-hand side of (2.5) with respect to u_k [2]

$$
\begin{aligned}
0 &= 2Ru_k + \frac{\partial V^*(x_{k+1})}{\partial u_k} \\
&= 2Ru_k + g^T(x_k) \frac{\partial V^*(x_{k+1})}{\partial x_{k+1}}.
\end{aligned} \tag{2.7}
$$

Therefore, the optimal control law is formulated as

$$
\begin{aligned}
u_k^* &= \bar{u}(x_k) \\
&= -\frac{1}{2} R^{-1} g^T(x_k) \frac{\partial V^*(x_{k+1})}{\partial x_{k+1}}
\end{aligned} \tag{2.8}
$$

which is a state feedback control law $\bar{u}(x_k)$ with $\bar{u}(0) = 0$. Hence, the boundary condition $V(0) = 0$ in (2.2) and (2.3) is satisfied for $V(x_k)$, and V becomes a Lyapunov function; u_k^* is used to emphasize that u_k is optimal.

Moreover, if $\mathcal{H}(x_k, u_k)$ is a quadratic form in u_k and $R > 0$, then

$$\frac{\partial^2 \mathcal{H}(x_k, u_k)}{\partial u_k^2} > 0$$

holds as a sufficient condition such that optimal control law (2.8) (globally [51]) minimizes $\mathcal{H}(x_k, u_k)$ and the performance index (2.2) [67].

Substituting (2.8) into (2.4), we obtain

$$
\begin{aligned}
V^*(x_k) &= l(x_k) + \left(-\frac{1}{2} R^{-1} g^T(x_k) \frac{\partial V^*(x_{k+1})}{\partial x_{k+1}} \right)^T \\
&\quad \times R \left(-\frac{1}{2} R^{-1} g^T(x_k) \frac{\partial V^*(x_{k+1})}{\partial x_{k+1}} \right) + V^*(x_{k+1}) \qquad (2.9) \\
&= l(x_k) + V^*(x_{k+1}) + \frac{1}{4} \frac{\partial V^{*T}(x_{k+1})}{\partial x_{k+1}} g(x_k) R^{-1} g^T(x_k) \frac{\partial V^*(x_{k+1})}{\partial x_{k+1}}
\end{aligned}
$$

which can be rewritten as

$$
l(x_k) + V^*(x_{k+1}) - V^*(x_k) + \frac{1}{4} \frac{\partial V^{*T}(x_{k+1})}{\partial x_{k+1}} g(x_k) R^{-1} g^T(x_k) \frac{\partial V^*(x_{k+1})}{\partial x_{k+1}} = 0. \quad (2.10)
$$

Equation (2.10) is known as the discrete-time HJB equation [2]. Solving this partial-differential equation for $V^*(x_k)$ is not straightforward. This is one of the main drawbacks in discrete-time optimal control for nonlinear systems. To overcome this problem, we propose using inverse optimal control.

2.2 LYAPUNOV STABILITY

In order to establish stability, we recall important related properties.

DEFINITION 2.1: Radially Unbounded Function [49] A positive definite function $V(x_k)$ satisfying $V(x_k) \to \infty$ as $\|x_k\| \to \infty$ is said to be radially unbounded.

DEFINITION 2.2: Decrescent Function [49] A function $V : \mathbb{R}^n \to \mathbb{R}$ is said to be decrescent if there is a positive definite function β such that the following inequality holds:

$$V(x_k) \leq \beta(\|x_k\|), \qquad \forall k \geq 0.$$

Theorem 2.1: Global Asymptotic Stability [61]

The equilibrium point $x_k = 0$ of (2.1) is globally asymptotically stable if there exists

a function $V : \mathbb{R}^n \to \mathbb{R}$ such that (i) V is a positive definite function, decrescent

and radially unbounded, and (ii) $-\Delta V(x_k, u_k)$ is a positive definite function, where

$\Delta V(x_k, u_k) = V(x_{k+1}) - V(x_k)$. ∎

Theorem 2.2: Exponential Stability [131]

Suppose that there exists a positive definite function $V : \mathbb{R}^n \to \mathbb{R}$ and constants

c_1, c_2, $c_3 > 0$ and $p > 1$ such that

$$c_1 \|x\|^p \leq V(x_k) \leq c_2 \|x\|^p \tag{2.11}$$

$$\Delta V(x_k) \leq -c_3 \|x\|^p, \qquad \forall k \geq 0, \quad \forall x \in \mathbb{R}^n. \tag{2.12}$$

Then $x_k = 0$ is an exponentially stable equilibrium for system (2.1). ∎

Clearly, exponential stability implies asymptotic stability. The converse is, how-

ever, not true.

Due to the fact that the inverse optimal control is based on a Lyapunov function,

we establish the following definitions.

DEFINITION 2.3: Control Lyapunov Function [7, 48] Let $V(x_k)$ be a radially

unbounded function, with $V(x_k) > 0$, $\forall x_k \neq 0$ and $V(0) = 0$. If for any $x_k \in \mathbb{R}^n$ there

exist real values u_k such that

$$\Delta V(x_k, u_k) < 0$$

where we define the Lyapunov difference as $\Delta V(x_k, u_k) = V\left(f(x_k) + g(x_k)u_k\right) -$

$V(x_k)$, then $V(\cdot)$ is said to be a discrete-time control Lyapunov function (CLF) for

system (2.1).

Assumption 1 *Let assume that $x = 0$ is an equilibrium point for (2.1), and that there*

exists a control Lyapunov function $V(x_k)$ such that

$$\alpha_1(\|x_k\|) \leq V(x_k) \leq \alpha_2(\|x_k\|) \tag{2.13}$$

$$\Delta V(x_k, u_k) \leq -\alpha_3(\|x_k\|) \tag{2.14}$$

where α_1, α_2, and α_3 are class $\mathcal{K}_\infty$ functions[3] and $\|\cdot\|$ denotes the usual Euclidean

norm. Then, the origin of the system is an asymptotically stable equilibrium point by

means of u_k as input.

The existence of this CLF is guaranteed by a converse theorem of the Lyapunov

stability theory [10].

As a special case, the calculus of class $\mathcal{K}_\infty-$ functions in (2.13) simplifies when

they take the special form $\alpha_i(r) = \kappa_i r^c$, $\kappa_i > 0$, $c = 2$, and $i = 1, 2$. In particular, for

---

[3]α_i, $i = 1, 2, 3$ belong to class $\mathcal{K}_\infty$ functions because later we will select a radially unbounded function

$V(x_k)$.

a quadratic positive definite function $V(x_k) = \frac{1}{2}x_k^T P x_k$, with P a positive definite and symmetric matrix, inequality (2.13) results in

$$\lambda_{min}(P)\|x\|^2 \leq x_k^T P x_k \leq \lambda_{max}(P)\|x\|^2 \tag{2.15}$$

where $\lambda_{min}(P)$ is the minimum eigenvalue of matrix P and $\lambda_{max}(P)$ is the maximum eigenvalue of matrix P.

2.3   ROBUST STABILITY ANALYSIS

This section reviews stability results for disturbed nonlinear systems, for which non-vanishing disturbances are considered. We can no longer study the stability of the origin as an equilibrium point, nor should we expect the solution of the disturbed system to approach the origin as $k \to \infty$. The best we can hope for is that if the disturbance is small in some sense, then the system solution will be ultimately bounded by a small bound [49], which conducts to the concept of ultimate boundedness.

DEFINITION 2.4: Ultimate Bound [26, 49] The solutions of (2.1) with $u_k = 0$ are said to be uniformly ultimately bounded if there exist positive constants b and c, and for every $a \in (0, c)$ there is a positive constant $T = T(a)$, such that

$$\|x_0\| < a \Rightarrow \|x_k\| \leq b, \quad \forall k \geq k_0 + T \tag{2.16}$$

where k_0 is the initial time instant. They are said to be globally uniformly ultimately bounded if (2.16) holds for arbitrarily large a. The constant b in (2.16) is known as the *ultimate bound*.

DEFINITION 2.5: $\mathscr{K}$ **and** $\mathscr{K}_\infty$ **functions [104]** A function $\gamma : \mathbb{R}_{\geq 0} \to \mathbb{R}_{\geq 0}$ is a $\mathscr{K}$–function if it is continuous, strictly increasing and $\gamma(0) = 0$; it is a $\mathscr{K}_\infty$–function if it is a $\mathscr{K}$–function and also $\gamma(s) \to \infty$ as $s \to \infty$; and it is a positive definite function if $\gamma(s) > 0$ for all $s > 0$, and $\gamma(0) = 0$.

DEFINITION 2.6: $\mathscr{K}\mathscr{L}$**-function [104]** A function $\beta : \mathbb{R}_{\geq 0} \times \mathbb{R}_{\geq 0} \to \mathbb{R}_{\geq 0}$ is a $\mathscr{K}\mathscr{L}$-function if, for each fixed $t \geq 0$, the function $\beta(\cdot, t)$ is a $\mathscr{K}$–function, and for each fixed $s \geq 0$, the function $\beta(s, \cdot)$ is decreasing and $\beta(s, t) \to 0$ as $t \to \infty$. $\mathbb{R}_{\geq 0}$ means nonnegative real numbers.

DEFINITION 2.7: BIBS [104] System (2.1) is uniformly bounded-input bounded-state (BIBS) stable with respect to u_k, if bounded initial states and inputs produce uniformly bounded trajectories.

DEFINITION 2.8: ISS Property [104, 76] System (2.1) is (globally) input-to-state stable (ISS) with respect to u_k if there exist a $\mathscr{K}\mathscr{L}$– function β and a $\mathscr{K}$– function γ such that, for each input $u \in l_\infty^m$ and each $x_0 \in \mathbb{R}^n$, it holds that the solution of (2.1) satisfies

$$\|x_k\| \leq \beta(\|x_0\|, k) + \gamma\left(\sup_{\tau \in [k_0, \infty)} \|u_\tau\| \right) \tag{2.17}$$

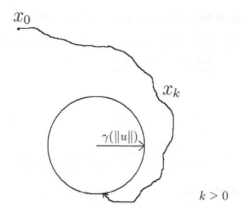

FIGURE 2.1 System trajectories with the ISS property.

where $\sup_{\tau \in [k_0, \infty)} \{\|u_\tau\| : \tau \in \mathbb{Z}^+\} < \infty$, which is denoted by $u \in \ell_\infty^m$.

Thus, system (2.1) is said to be ISS if property (2.17) is satisfied [57]. The interpretation of (2.17) is the following: for a bounded input u, the system solution remains in the ball of radius $\beta(\|x_0\|, k) + \gamma\left(\sup_{\tau \in [k_0, \infty)} \|u_\tau\|\right)$. Furthermore, as k increases, all trajectories approach the ball of radius $\gamma\left(\sup_{\tau \in [k_0, \infty)} \|u_\tau\|\right)$ (i.e., all trajectories will be ultimately bounded with ultimate bound γ). Because γ is of class $\mathcal{K}$, this ball is a small neighborhood of the origin whenever $\|u\|$ is small (see Figure 2.1). ISS is used to analyze the stability of the solutions for disturbed nonlinear systems. The ISS property captures the notion of BIBS stability.

DEFINITION 2.9: Asymptotic Gain Property [104] System (2.1) is said to have the $\mathscr{K}$ – asymptotic gain if there exists some $\gamma \in \mathscr{K}$ such that

$$\lim_{k \to \infty} \|x_k(x_0, u)\| \leq \lim_{k \to \infty} \gamma(\|u_k\|) \tag{2.18}$$

for all $x_0 \in \mathbb{R}^n$.

Theorem 2.3: ISS System [104]

Consider system (2.1). The following are equivalent:

(1) It is ISS.

(2) It is BIBS and it admits $\mathscr{K}$ – asymptotic gain.   ∎

Let ℓ_d be the Lipschitz constant such that for all β_1 and β_2 in some bounded

neighborhood of (x_k, u_k), the Lyapunov function $V(x_k)$ satisfies the condition [120]

$$\|V(\beta_1) - V(\beta_2)\| \leq \ell_d \|\beta_1 - \beta_2\|, \quad \ell_d > 0. \tag{2.19}$$

DEFINITION 2.10: ISS – Lyapunov Function [104] A continuous function V on

$\mathbb{R}^n$ is called an ISS–Lyapunov function for system (2.1) if

$$\alpha_1(\|x_k\|) \leq V(x_k) \leq \alpha_2(\|x_k\|) \tag{2.20}$$

holds for some α_1, $\alpha_2 \in \mathscr{K}_\infty$, and

$$V(f(x_k, u_k)) - V(x_k) \leq -\alpha_3(\|x_k\|) + \sigma(\|u_k\|) \tag{2.21}$$

for some $\alpha_3 \in \mathscr{K}_\infty$, $\sigma \in \mathscr{K}$. A smooth ISS–Lyapunov function is one which is smooth.

Note that if $V(x_k)$ is an ISS–Lyapunov function for (2.1), then $V(x_k)$ is a DT Lyapunov function for the 0-input system $x_{k+1} = f(x_k) + g(x_k)0$.

PROPOSITION 2.1

If system (2.1) admits an ISS–Lyapunov function, then it is ISS [104].

Now, consider the disturbed system

$$x_{k+1} = f(x_k) + g(x_k)u_k + d_k, \qquad x_0 = x(0) \tag{2.22}$$

where $x_k \in \mathbb{R}^n$ is the state of the system at time $k \in \mathbb{Z}^+$, $u_k \in \mathbb{R}^m$ is the control, $d_k \in \mathbb{R}^n$ is the disturbance term, $f : \mathbb{R}^n \to \mathbb{R}^n$ and $g : \mathbb{R}^n \to \mathbb{R}^{n \times m}$ are smooth mappings, $f(0) = 0$. The perturbation term d_k could result from modeling errors, aging, or uncertainties and disturbances which exist for any realistic problem [49].

DEFINITION 2.11: ISS–CLF Function A smooth positive definite radially unbounded function $V : \mathbb{R}^n \to \mathbb{R}$ is said to be an ISS–CLF for system (2.22) if there exists a class $\mathscr{K}_\infty$ function ρ such that the following inequalities hold $\forall x \neq 0$ and $\forall d \in \mathbb{R}^n$:

$$\alpha_1(\|x_k\|) \leq V(x_k) \leq \alpha_2(\|x_k\|) \tag{2.23}$$

for some α_1, $\alpha_2 \in \mathscr{K}_\infty$, and

$$\|x_k\| \geq \rho(\|d_k\|) \Rightarrow \inf_{u_k \in \mathbb{R}^m} \Delta V_d(x_k, d_k) < -\alpha_3(\|x_k\|) \tag{2.24}$$

where $\Delta V_d(x_k, d_k) := V(x_{k+1}) - V(x_k)$ and $\alpha_3 \in \mathcal{K}_\infty$.

REMARK 2.1 The connection between the existence of a Lyapunov function and the input-to-state stability is that an estimate of the gain function γ in (2.17) is $\gamma = \alpha_1^{-1} \circ \alpha_2 \circ \rho$, where $\circ$ means composition[4] of functions with α_1 and α_2 as defined in (2.23) [58]. ∎

Note that if $V(x_k)$ is an ISS–control Lyapunov function for (2.22), then $V(x_k)$ is a control Lyapunov function for the 0-disturbance system $x_{k+1} = f(x_k) + g(x_k) u_k$.

PROPOSITION 2.2: ISS–CLF System

If system (2.22) admits an ISS–CLF, then it is ISS.

2.3.1 OPTIMAL CONTROL FOR DISTURBED SYSTEMS

For disturbed discrete-time nonlinear system (2.22), the Bellman equation becomes the Isaacs equation described by

$$V(x_k) = \min_{u_k} \left\{ l(x_k) + u_k^T R(x_k) u_k + V(x_k, u_k, d_k) \right\} \tag{2.25}$$

---

[4] $\alpha_1(\cdot) \circ \alpha_2(\cdot) = \alpha_1(\alpha_2(\cdot))$.

and the Hamilton–Jacobi–Isaacs (HJI) equation associated with system (2.22) and cost functional (2.2) is

$$
\begin{aligned}
0 &= \inf_u \sup_{d \in \mathscr{D}} \left\{ l(x_k) + u_k^T R(x_k) u_k + V(x_{k+1}) - V(x_k) \right\} \\
&= \inf_u \sup_{d \in \mathscr{D}} \left\{ l(x_k) + u_k^T R(x_k) u_k + V(x_k, u_k, d_k) - V(x_k) \right\} \quad (2.26)
\end{aligned}
$$

where $\mathscr{D}$ is the set of locally bounded functions, and function $V(x_k)$ is unknown. However, finding a solution of HJI equation (2.26) for $V(x_k)$ with (2.8) is the main drawback of the robust optimal control; this solution may not exist or may be pretty difficult to solve [36]. Note that $V(x_{k+1})$ in (2.26) is a function of the disturbance term d_k.

2.4 PASSIVITY

Let us consider a nonlinear affine system and an output given as

$$
x_{k+1} = f(x_k) + g(x_k) u_k, \qquad x_0 = x(0) \quad (2.27)
$$

$$
y_k = h(x_k) + J(x_k) u_k \quad (2.28)
$$

where $x_k \in \mathbb{R}^n$ is the state of the system at time k, output $y_k \in \mathbb{R}^m$; $h(x_k) : \mathbb{R}^n \to \mathbb{R}^m$, and $J(x_k) : \mathbb{R}^n \to \mathbb{R}^{m \times m}$ are smooth mappings. We assume $h(0) = 0$.

We present definitions, sufficient conditions, and key results, which help us to solve the inverse optimal control via passivity as follows.

DEFINITION 2.12: Passivity [17] System (2.27)–(2.28) is said to be passive if there exists a nonnegative function $V(x_k)$, called the storage function, such that for all u_k,

$$V(x_{k+1}) - V(x_k) \leq y_k^T u_k \qquad (2.29)$$

where $(\cdot)^T$ denotes transpose.

This storage function may be selected as a CLF candidate if it is a positive definite function [121]. It is worth noting that the output which renders the system passive is not in general the variable we wish to control, and it is used only for control synthesis.

DEFINITION 2.13: Zero–State Observable System [18] A system (2.27)–(2.28) is locally zero-state observable (respectively locally zero-state detectable) if there exists a neighborhood $\mathscr{Z}$ of $x_k = 0$ in $\mathbb{R}^n$ such that for all $x_0 \in \mathscr{Z}$

$$y_k|_{u_k=0} = h(\phi(k,x_0,0)) = 0 \quad \forall k \implies x_k = 0 \quad \left(respectively \ \lim_{k \to \infty} \phi(k,x_0,0) = 0 \right)$$

where $\phi(k,x_0,0) = f^k(x_k)$ is the trajectory of the unforced dynamics $x_{k+1} = f(x_k)$ with initial condition x_0. If $\mathscr{Z} = \mathbb{R}^n$, the system is zero-state observable (respectively zero-state detectable).

Additionally, the following definition is introduced.

DEFINITION 2.14: Feedback Passive System System (2.27)–(2.28) is said to be feedback passive if there exists a passifying law

$$u_k = \alpha(x_k) + v_k, \qquad \alpha, v \in \mathbb{R}^m \qquad (2.30)$$

with a smooth function $\alpha(x_k)$ and a storage function $V(x)$, such that system (2.27)

with (2.30), described by

$$x_{k+1} = \bar{f}(x_k) + g(x_k) v_k, \qquad x_0 = x(0) \qquad (2.31)$$

and output

$$\bar{y}_k = \bar{h}(x_k) + J(x_k) v_k \qquad (2.32)$$

satisfies relation (2.29) with v_k as the new input, where $\bar{f}(x_k) = f(x_k) + g(x_k) \alpha(x_k)$

and $\bar{h} : \mathbb{R}^n \to \mathbb{R}^m$ is a smooth mapping, which will be defined later, with $\bar{h}(0) = 0$.

Roughly speaking, to render system (2.27) feedback passive can be summarized as

determining a passivation law u_k and an output $\bar{y}_k$, such that relation (2.29) is satisfied

with respect to the new input v_k.

2.5   NEURAL IDENTIFICATION

Analysis of nonlinear systems requires of a lot of effort, since parameters are difficult

to obtain [38]. Hence, to synthesize a controller based on the plant model which has

uncertainties is not practical.

For realistic situations, a control based on a plant model cannot perform as de-

sired, due to internal and external disturbances, uncertain parameters, or unmodeled

dynamics [38]. This fact motivates the need to derive a model based on RHONN to

identify the dynamics of the plant.

We analyze a general class of systems which are affine in the control with distur-

bance terms as in [22]; the same structure is assumed for the neural network.

2.5.1 NONLINEAR SYSTEMS

Consider a class of discrete-time disturbed nonlinear system

$$\chi_{k+1} = \bar{f}(\chi_k) + \bar{g}(\chi_k)u_k + \Gamma_k \tag{2.33}$$

where $\chi_k \in R^n$ is the system state at time k, $\Gamma_k \in R^n$ is an unknown and bounded perturbation term representing modeling errors, uncertain parameters, and disturbances; $\bar{f} : \mathbb{R}^n \to \mathbb{R}^n$ and $\bar{g} : \mathbb{R}^n \to \mathbb{R}^{n \times m}$ are smooth mappings. Without loss of generality, $\chi_k = 0$ is an equilibrium point for (2.33). We assume $\bar{f}(0) = 0$ and $rank\{\bar{g}(\chi_k)\} = m$ $\forall \chi_k \neq 0$.

2.5.2 DISCRETE-TIME RECURRENT HIGH ORDER NEURAL NETWORK

To identify system (2.33), let us consider the following discrete-time RHONN proposed in [116]:

$$x_{i,k+1} = w_{i,k}^T \rho_i(x_k, u_k) \tag{2.34}$$

where $x_k = [x_{1,k} \, x_{2,k} \, \ldots \, x_{n,k}]^T$, x_i is the state of the i-th neuron which identifies the i-th component of state vector χ_k in (2.33), $i = 1, \ldots, n$; w_i is the respective on-line adapted weight vector, and $u_k = [u_{1,k} \, u_{2,k} \, \ldots \, u_{m,k}]^T$ is the input vector to the neural network; ρ_i is an L_p dimensional vector defined as

$$\rho_i(x_k, u_k) = \begin{bmatrix} \rho_{i_1} \\ \rho_{i_2} \\ \vdots \\ \rho_{i_{L_p}} \end{bmatrix} = \begin{bmatrix} \prod_{\ell \in I_1} z_{i_\ell}^{d_{i_\ell}(1)} \\ \prod_{\ell \in I_2} z_{i_\ell}^{d_{i_\ell}(2)} \\ \vdots \\ \prod_{\ell \in I_{L_p}} z_{i_\ell}^{d_{i_\ell}(L_p)} \end{bmatrix} \tag{2.35}$$

where d_{i_ℓ} are nonnegative integers, L_p is the respective number p of high-order connections, and $\{I_1, I_2, \ldots, I_{L_p}\}$ is a collection of non-ordered subsets of $\{1, 2, \ldots, n+m\}$. Z_i is a vector defined as

$$
Z_i = \begin{bmatrix} Z_{i_1} \\ \vdots \\ Z_{i_n} \\ Z_{i_{n+1}} \\ \vdots \\ Z_{i_{n+m}} \end{bmatrix} = \begin{bmatrix} S(x_{1,k}) \\ \vdots \\ S(x_{n,k}) \\ u_{1,k} \\ \vdots \\ u_{m,k} \end{bmatrix}
$$

where the sigmoid function $S(\cdot)$ is defined by

$$
S(x) = \frac{\alpha_i}{1 + e^{-\beta_i x}} - \gamma_i \tag{2.36}
$$

with $S(\cdot) \in [-\gamma_i, \alpha_i - \gamma_i]$; α_i, β_i, and γ_i are positive constants. An i-th RHONN scheme is depicted in Figure 2.2.

We propose the following modification of the discrete-time RHONN (2.34) for the system described by (2.33) [92]:

- Neural weights associated with the control inputs could be fixed (w_i') to ensure controllability of the identifier.

Based on this modification and using the structure of system (2.33), we propose the following neural network model:

$$
x_{i,k+1} = w_{i,k}^T \rho_i(x_k) + {w_i'}^T \psi_i(x_k, u_k) \tag{2.37}
$$

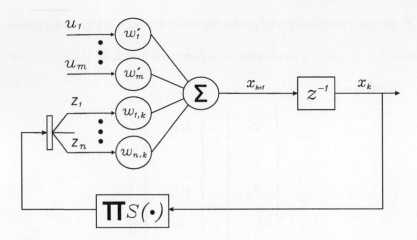

FIGURE 2.2   RHONN scheme.

in order to identify (2.33), where x_i is the i-th neuron state; $w_{i,k}$ is the on-line adjustable weight vector, and w_i' is the fixed weight vector; ψ denotes a function of x or u corresponding to the plant structure (2.33) or external inputs to the network, respectively. Vector ρ_i in (2.37) is like (2.35); however, Z_i is redefined as

$$
Z_i = \begin{bmatrix} Z_{i_1} \\ \vdots \\ Z_{i_n} \end{bmatrix} = \begin{bmatrix} S(x_{1,k}) \\ \vdots \\ S(x_{n,k}) \end{bmatrix}
$$

The on-line adjustable weight vector $w_{i,k}$ is defined as

$$
w_{i,k} = \begin{bmatrix} w_{i1,k} & \cdots & w_{iL_p,k} \end{bmatrix}^T.
$$

REMARK 2.2 It is worth noting that (2.37) does not consider the disturbance term (Γ_k) because the RHONN weights are adjusted on-line, and hence the RHONN identifies the dynamics of the nonlinear system, which includes the disturbance effects.

■

2.5.2.1   RHONN Models

From results presented in [113], we can assume that there exists a RHONN which

models (2.33); thereby, plant model (2.33) can be described by

$$\chi_{k+1} = W_k^* \rho(\chi_k) + W'^* \psi(\chi_k, u_k) + v_k \tag{2.38}$$

where $W_k^* = [w_{1,k}^{*T}\ w_{2,k}^{*T}\ \cdots\ w_{n,k}^{*T}]^T$ and $W'^* = [w_1'^{*T}\ w_2'^{*T}\ \cdots\ w_n'^{*T}]^T$ are the optimal

unknown weight matrices, and the modeling error v_k is given by

$$v_k = \bar{f}(\chi_k) + \bar{g}(\chi_k)\, u_k + \Gamma_k - W_k^* \rho(\chi_k) - W'^* \psi(\chi_k, u_k).$$

The modeling error term v_k can be rendered arbitrarily small by selecting appropriately

the number L_p of high-order connections [113]. The ideal weight matrices W_k^* and

W'^* are artificial quantities required for analytical purpose. In general, it is assumed

that this vector exists and is constant but unknown. Optimal unknown weights $w_{i,k}^*$

will be approximate by the on-line adjustable ones $w_{i,k}$ [116].

For neural identification of (2.33), two possible models for (2.37) can be used.

- Parallel model

$$x_{i,k+1} = w_{i,k}^T \rho_i(x_k) + w_i'^T \psi_i(x_k, u_k). \tag{2.39}$$

- Series–parallel model

$$x_{i,k+1} = w_{i,k}^T \rho_i(\chi_k) + w_i'^T \psi_i(\chi_k, u_k). \tag{2.40}$$

2.5.2.2 On-line Learning Law

For the RHONN weights on-line learning, we use an EKF [115]. The weights become

the states to be estimated [40, 122]; the main objective of the EKF is to find the optimal

values for the weight vector $w_{i,k}$ such that the prediction error is minimized. The EKF

solution to the training problem is given by the following recursion:

$$
\begin{aligned}
M_{i,k} &= [R_{i,k} + H_{i,k}^T P_{i,k} H_{i,k}]^{-1} \\
K_{i,k} &= P_{i,k} H_{i,k} M_{i,k} \\
w_{i,k+1} &= w_{i,k} + \eta_i K_{i,k} e_{i,k} \\
P_{i,k+1} &= P_{i,k} - K_{i,k} H_{i,k}^T P_{i,k} + Q_{i,k}
\end{aligned}
\tag{2.41}
$$

where vector $w_{i,k}$ represents the estimate of the i-th weight (state) of the i-th neuron

at update step k. This estimate is a function of the Kalman gain K_i and the neural

identification error $e_{i,k} = \chi_{i,k} - x_{i,k}$, where χ_i is the plant state and x_i is the RHONN

state. The Kalman gain is a function of the approximate error covariance matrix P_i,

a matrix of derivatives of the network's outputs with respect to all trainable weight

parameters H_i as follows:

$$
H_{i,k} = \left[\frac{\partial x_{i,k}}{\partial w_{i,k}} \right]^T
\tag{2.42}
$$

and a global scaling matrix M_i. Here, Q_i is the covariance matrix of the process noise

and R_i is the measurement noise covariance matrix. As an additional parameter we

introduce the rate learning η_i such that $0 \leq \eta_i \leq 1$. Usually P_i, Q_i, and R_i are initialized

as diagonal matrices, with entries $P_i(0)$, $Q_i(0)$, and $R_i(0)$, respectively. We set Q_i and

R_i fixed. During training, the values of H_i, K_i, and P_i are ensured to be bounded [116].

Theorem 2.4: Identification via RHONN [116]

The RHONN (2.34) trained with the EKF based algorithm (2.41) to identify the nonlinear plant (2.33) ensures that the neural identification error $e_{i,k}$ is semiglobally uniformly ultimately bounded (SGUUB); moreover, the RHONN weights remain bounded. ∎

2.5.3 DISCRETE-TIME RECURRENT MULTILAYER PERCEPTRON

In this section for the neural identification, the recurrent multi-layer perceptron (RMLP) is selected; then the neural model structure definition reduces to dealing with the following issues: 1) determining the inputs to the neural network (NN) and 2) determining the internal architecture of the NN.

The selected structure of the neural scheme is the Neural Network Autoregresive eXternal (NNARX) input [90]; the output vector for the Artificial Neural Network (ANN) is defined as the regression vector of an Autoregresive eXternal (ARX) input linear model structure [117].

It is common to consider a general discrete-time nonlinear system $y_{k+1} = \bar{f}(y_k, u_k)$; however, for many control applications it is preferred to express the model in an affine form, which can be represented by the following equation:

$$y_{k+1} = f\left(y_k, y_{k-1}, \dots, y_{k-q+1}\right) + g u_k \tag{2.43}$$

where q is the dimension of the state space and g is the input matrix. A nonlinear mapping f exists, for which the value of the output y_{k+1} is uniquely defined in terms of its past values $y_k, \dots, y_{k-q+1}$ and the present value of the input u_k.

Considering that it is possible to define

$$\phi_k = \left[y_k, \ldots, y_{k-q+1}\right]^T \tag{2.44}$$

which is similar to the regression vector of an ARX linear model structure [90], then

the nonlinear mapping f can be approximated by a neural network defined as

$$y_{k+1} = S(\phi_k, w^*) + w'^* u_k + \varepsilon \tag{2.45}$$

where $S(\cdot)$ contains the required sigmoidal terms. It is assumed that there exist ideal

weight vectors w^* and w'^* such that the neural modeling error ε is minimized on a

compact set [42]. The ideal weight vectors w^* and w'^* are artificial quantities required

only for analytical purposes [53]. In general, it is assumed that these vectors exist and

are constant but unknown. Let us define their estimates as w and w', respectively; then

the estimation errors are defined as

$$\widetilde{w}_k = w^* - w_k \tag{2.46}$$

$$\widetilde{w'} = w'^* - w'. \tag{2.47}$$

The neural network (2.45) can be implemented on a predictor form as

$$\widehat{y}_{k+1} = w_k^T S(\phi_k) + w' u_k \tag{2.48}$$

where w is the vector containing the adjustable parameters and w' is a fixed weight

vector for inputs, which is used to ensure controllability of the neural model [94], and

e_k is the prediction error defined as

$$e_k = y_k - \widehat{y}_k \tag{2.49}$$

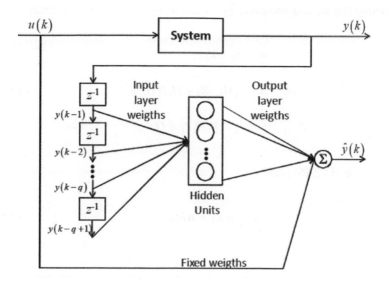

FIGURE 2.3   Neural network structure.

which includes all the effects produced by the neural network approximation, exter-

nal disturbances, and plant parameter variations. Hence (2.48) constitutes an RMLP

neural network.

It is important to note that since [53], neural identification has been discussed

in many publications ([5] ,[66], [90] and references therein), therefore it is omitted.

However, the neural model (2.48) can be seen as a special case of the neural identifier

presented in [5]. The neural network structure is presented in Figure 2.3.

The RMLP used in this book contains sigmoid units only in the hidden layer; the

output layer is a linear one.

The dynamics of (2.49) can be expressed as

$$e_{k+1} = \widetilde{w}_k z\left(x_k, u_k\right) + \widetilde{w}' u_k + \varepsilon. \tag{2.50}$$

Considering that w_k is determined by the following EKF-based algorithm

$$w_{k+1} = w_k + K_k e_k$$

$$K_k = P_k H_k^T \left[R_k + H_k P_k H_k^T \right]^{-1} \qquad (2.51)$$

$$P_{k+1} = P_k - K_k H_k P_k + Q_k$$

then the dynamic of (2.46) results in

$$\widetilde{w}_{k+1} = \widetilde{w}_k - \eta K_k e_k. \qquad (2.52)$$

Theorem 2.5: Identification via NNARX

The NNARX (2.48) trained with the modified EKF-based algorithm (2.51) to identify the nonlinear plant (2.45) ensures that the prediction error (2.49) is SGUUB; moreover, the RHONN weights remain bounded.   ■

PROOF

Consider the Lyapunov function candidate.

$$V_k = \widetilde{w}_k^T P_k \widetilde{w}_k + e_k^2. \qquad (2.53)$$

Its first difference can be written as

$$\Delta V_k = V_{k+1} - V_k$$

$$= \widetilde{w}_{k+1}^T P_{k+1} \widetilde{w}_{k+1} + e_{k+1}^2 - \widetilde{w}_k^T P_k \widetilde{w}_k - e_k^2. \qquad (2.54)$$

Using (2.50) and (2.52) in (2.54)

$$\Delta V_k = [\widetilde{w}_k - \eta K_k e_k]^T [P_k - A_k] [\widetilde{w}_k - \eta K_k e_k]$$
$$+ \left(\widetilde{w}_k z(x_k, u_k) + \widetilde{w}' u_k + \varepsilon\right)^2 - \widetilde{w}_k^T P_k \widetilde{w}_k - e_k^2 \qquad (2.55)$$

with $A_k = K_k H_k^T P_k - Q_k$; then, (2.55) can be expressed as

$$\Delta V_k = \widetilde{w}_k^T P_k \widetilde{w}_k - 2\eta \widetilde{w}_k^T P_k K_k e_k + \eta^2 e_k^2 K_k^T P_k K_k - \widetilde{w}_k^T A_k \widetilde{w}_k$$
$$+ 2\eta \widetilde{w}_k^T A_k K_k e_k - \eta^2 e_k^2 K_k^T A_k K_k + \left(\widetilde{w}_k z(x_k, u_k)\right)^2$$
$$+ 2\widetilde{w}_k z(x_k, u_k) \widetilde{w}' u_k + \left(\widetilde{w}' u_k\right)^2 + 2\widetilde{w}_k z(x_k, u_k) \varepsilon$$
$$+ 2\widetilde{w}' u_k \varepsilon + \varepsilon^2 - \widetilde{w}_k^T P_k \widetilde{w}_k - e_k^2. \qquad (2.56)$$

Using the following inequalities

$$X^T X + Y^T Y \geq 2X^T Y$$
$$X^T X + Y^T Y \geq -2X^T Y$$
$$-\lambda_{\min}(P) X^2 \geq -X^T P X \geq -\lambda_{\max}(P) X^2 \qquad (2.57)$$

which are valid $\forall X, Y \in \Re^n, \forall P \in \mathbb{R}^{n \times n}, P = P^T > 0$, then (2.56), can be rewritten as

$$\Delta V_k \leq \eta^2 e_k^2 K_k^T P_k K_k - \widetilde{w}_k^T A_k \widetilde{w}_k - \eta^2 e_k^2 K_k^T A_k K_k$$
$$+ 2\left(\widetilde{w}' u_k\right)^2 - e_k^2 + e_k^2 K_k^T P_k P_k^T K_k + 2\eta^2 \widetilde{w}_k^T \widetilde{w}_k$$
$$+ e_k^2 K_k^T A_k A^T K_k + 3\left(\widetilde{w}_k z(x_k, u_k)\right)^2 + 2\varepsilon^2$$
$$\leq \eta^2 e_k^2 \|K_k\|^2 \lambda_{\max}(P_k) - \lambda_{\min}(A_k) \|\widetilde{w}_k\|^2 - \eta^2 e_k^2 \|K_k\|^2 \lambda_{\min}(A_k)$$
$$+ 2\left(\widetilde{w}' u_k\right)^2 - e_k^2 + e_k^2 \|K_k\|^2 \lambda_{\max}^2(P_k) + 2\eta^2 \|\widetilde{w}_k\|^2$$
$$+ e_k^2 \|K_k\|^2 \lambda_{\max}^2(A_k) + 3\|\widetilde{w}_k\|^2 \|z(x_k, u_k)\|^2 + 2\varepsilon^2. \qquad (2.58)$$

Defining

$$E_k = \lambda_{\min}(A_k) - 3\|z(x_k, u_k)\|^2 - 2\eta^2$$

$$F_k = \eta^2 \|K_k\|^2 \lambda_{\min}(A_k) - \eta^2 \|K_k\|^2 \lambda_{\max}(P_k)$$

$$- \|K_k\|^2 \lambda_{\max}^2(P_k) - \|K_k\|^2 \lambda_{\max}^2(A_k) + 1$$

$$G_k = 2\varepsilon^2 + 2(\widetilde{w}' u_k)^2$$

and selecting η, Q_k, and R_k, such that $E_k > 0$ and $F_k > 0$, $\forall k$, then (2.58) can be expressed as

$$\Delta V_k \leq -\|\widetilde{w}_k\|^2 E_k - e_k^2 F_k + G_k.$$

Hence, $\Delta V_i(k) < 0$ when

$$\|\widetilde{w}_k\| > \sqrt{\frac{G_k}{E_k}} \equiv \kappa_1$$

or

$$|e_k| > \sqrt{\frac{G_k}{F_k}} \equiv \kappa_2.$$

Therefore the solution of (2.50) and (2.52) is SGUUB; hence, the estimation error and the RHONN weights are SGUUB [50]. ∎

3 Inverse Optimal Control:

A Passivity Approach

This chapter deals with inverse optimal control via passivity for both stabilization and trajectory tracking. In Section 3.1, a stabilizing inverse optimal control is synthesized. In Section 3.2, trajectory tracking is presented by modifying the proposed CLF such that it has a global minimum along the desired trajectory. Examples illustrate the proposed control scheme applicability. Finally, Section 3.3 presents an inverse optimal control scheme for nonlinear positive systems.

3.1 INVERSE OPTIMAL CONTROL VIA PASSIVITY

In this section, we proceed to develop an inverse optimal control law for system (2.27), which can be globally asymptotically stabilized by the output feedback $u_k = -y_k$. It is worth mentioning that the output with respect to which the system is rendered passive could not be the variable which we wish to control. The passive output will only be a preliminary step for control synthesis; additionally, we need to ensure that the output variables, which we want to control, behave as desired.

Let us state the conditions to achieve inverse optimality via passivation in the following theorem, for which the storage function is used as a CLF.

Theorem 3.1

Let an affine discrete-time nonlinear system (2.27) with input (2.30) and output (2.32)

be zero-state detectable. Consider passivity condition (2.29) with a CLF candidate as

$V(x_k) = \frac{1}{2} x_k^T P x_k$, $P = P^T > 0$, and a control input as defined in (2.30) with $\alpha(x_k)$

given by

$$\alpha(x_k) = -(I_m + J(x_k))^{-1} h(x_k) \tag{3.1}$$

where I_m is the $m \times m$ identity matrix, $(\cdot)^{-1}$ denotes inverse, and $h(x_k)$ and $J(x_k)$ are

defined as

$$h(x_k) = g^T(x_k) P f(x_k) \tag{3.2}$$

and

$$J(x_k) = \frac{1}{2} g^T(x_k) P g(x_k). \tag{3.3}$$

If there exists P such that the following inequality holds

$$(f(x_k) + g(x_k) \alpha(x_k))^T P (f(x_k) + g(x_k) \alpha(x_k)) - x_k^T P x_k \leq 0, \tag{3.4}$$

then **(a)** system (2.27) with (2.30) and (2.32) is feedback passive for $\bar{y}_k$ in (2.32)

defined as $\bar{y}_k = \bar{h}(x_k) + J(x_k) v_k$, $\bar{h}(x_k) = g^T(x_k) P \bar{f}(x_k)$, and $\bar{f}(x_k) = f(x_k) +$

$g(x_k) \alpha(x_k)$; **(b)** system (2.27) with (2.30) is globally asymptotically stabilized at

the equilibrium point $x_k = 0$ by the output feedback $v_k = -\bar{y}_k$; and **(c)** with $V(x_k)$

as a CLF, control law (3.1) is inverse optimal in the sense that it minimizes the cost

functional

$$\mathscr{I} = \sum_{k=0}^{\infty} L(x_k, \alpha(x_k)) \tag{3.5}$$

where $L(x_k, \alpha(x_k)) = l(x_k) + \alpha^T(x_k)\alpha(x_k)$, $l(x_k) = -\frac{\bar{f}^T(x_k)P\bar{f}(x_k)-x_k^T P x_k}{2} \geq 0$, and

optimal value function $\mathscr{J}^* = V(x_0)$.   ∎

PROOF

(a) Passivation: System (2.27) with input (2.30) and output (2.32) must be rendered feedback passive such that the inequality $V(x_{k+1}) - V(x_k) \leq \bar{y}_k^T v_k$ is fulfilled, with v_k as new input and output $\bar{y}_k$ as defined in (2.32); hence, from this inequality we obtain

$$\frac{\left(f(x_k) + g(x_k)\alpha(x_k)\right)^T P \left(f(x_k) + g(x_k)\alpha(x_k)\right) - x_k^T P x_k}{2}$$
$$+ \frac{2\left(f(x_k) + g(x_k)\alpha(x_k)\right)^T P g(x_k) v_k + v_k^T g^T(x_k) P g(x_k) v_k}{2} \tag{3.6}$$
$$\leq \bar{h}^T(x_k) v_k + v_k^T J^T(x_k) v_k$$

where $\alpha(x_k)$ is defined by (3.1). Rewriting (3.6) in accordance with related terms for both sides of the equation, then

- From the first term of (3.6), we have

$$\left(f(x_k) + g(x_k)\alpha(x_k)\right)^T P \left(f(x_k) + g(x_k)\alpha(x_k)\right) - x_k^T P x_k \leq 0; \tag{3.7}$$

- $2\left(f(x_k) + g(x_k)\alpha(x_k)\right)^T P g(x_k) v_k = 2\bar{h}^T(x_k) v_k$, then we define $\bar{h}(x_k)$ as

$$\bar{h}(x_k) = g^T(x_k) P \left(f(x_k) + g(x_k)\alpha(x_k)\right)$$
$$= g^T(x_k) P \bar{f}(x_k); \tag{3.8}$$

- $v^T g^T(x_k) P g(x_k) v_k = 2 v_k^T J^T(x_k) v_k$, then we define $J(x_k)$ as

$$J(x_k) = \frac{1}{2} g^T(x_k) P g(x_k). \qquad (3.9)$$

From (3.7)–(3.9), we deduce that if there exists P such that (3.7) is satisfied, then

system (2.27) with (2.30) and (2.32) satisfies inequality (3.6) and therefore is feedback

passive with $V(x_k)$ as the storage function, and $\bar{h}(x_k)$ and $J(x_k)$ as defined in (3.8) and

(3.9), respectively.

(b) Stability: In order to establish asymptotic stability for the closed-loop system

(2.27), (2.30) by output feedback

$$
\begin{aligned}
v_k &= -\bar{y}_k \\
&= -(I_m + J(x_k))^{-1} \bar{h}(x_k) \qquad (3.10)
\end{aligned}
$$

as described in [69], let us consider the difference

$$
\begin{aligned}
\Delta V(x_k) &:= V(x_{k+1}) - V(x_k) \\
&= \frac{[\bar{f}(x_k) + g(x_k) v_k]^T P [\bar{f}(x_k) + g(x_k) v_k] - x_k^T P x_k}{2} \\
&= \frac{\bar{f}^T(x_k) P \bar{f}(x_k) - x_k^T P x_k}{2} + \bar{f}^T(x_k) P g(x_k) v_k + v_k^T \frac{1}{2} g^T(x_k) P g(x_k) v_k \\
&= \frac{\bar{f}^T(x_k) P \bar{f}(x_k) - x_k^T P x_k}{2} - \bar{h}^T(x_k) (I_m + J(x_k))^{-1} \bar{h}(x_k) \\
&\quad + \bar{h}^T(x_k) (I_m + J(x_k))^{-1} \frac{1}{2} g^T(x_k) P g(x_k) (I_m + J(x_k))^{-1} \bar{h}(x_k) \quad (3.11)
\end{aligned}
$$

and noting that $\frac{1}{2} g^T(x_k) P g(x_k) = (I_m + \frac{1}{2} g^T(x_k) P g(x_k)) - I_m = (I_m + J(x_k)) - I_m$, we obtain

$$
\begin{aligned}
\Delta V(x_k) &= \frac{\bar{f}^T(x_k)P\bar{f}(x_k) - x_k^T P x_k}{2} - \bar{h}^T(x_k)\,(I_m + J(x_k))^{-1}\,\bar{h}(x_k) \\[2mm]
&\quad + \bar{h}^T(x_k)\,(I_m + J(x_k))^{-1}\,[(I_m + J(x_k)) - I_m]\,(I_m + J(x_k))^{-1}\,\bar{h}(x_k) \\[2mm]
&= \frac{\bar{f}^T(x_k)P\bar{f}(x_k) - x_k^T P x_k}{2} - \bar{h}^T(x_k)\,(I_m + J(x_k))^{-2}\,\bar{h}(x_k). \qquad (3.12)
\end{aligned}
$$

From (3.10) we have $v_k = -(I_m + J(x_k))^{-1}\,\bar{h}(x_k)$, and under condition (3.7), the difference (3.12) results in

$$
\begin{aligned}
\Delta V(x_k) &= \frac{\bar{f}^T(x_k)P\bar{f}(x_k) - x_k^T P x_k}{2} - \|v_k\|^2 \\[2mm]
&< 0. \qquad\qquad\qquad\qquad\qquad\qquad\qquad\qquad (3.13)
\end{aligned}
$$

Due to the fact that $V(x_k) = \frac{1}{2}x_k^T P x_k$ is a radially unbounded function, the solution $x_k = 0$ of the closed-loop system (2.27), with (2.30) and output feedback (2.32), is globally asymptotically stable.

(c) **Optimality:** Control law (3.1) is established to be an inverse optimal law since (i) it stabilizes system (2.27) according to (b); and (ii) it minimizes cost functional (3.5).

To fulfill (ii), control law (3.1) minimizes (3.5) due to the fact that this control law is obtained from the Hamiltonian (2.5) by calculating $\frac{\partial \mathcal{H}(x_k, u_k)}{\partial u_k} = 0$ and considering

$L(x_k, \alpha(x_k))$ as defined in (3.5), that is,

$$
\begin{aligned}
0 &= \min_{\alpha(x_k)} \{L(x_k, \alpha(x_k)) + V(x_{k+1}) - V(x_k)\} \\
&= \min_{\alpha(x_k)} \{l(x_k) - y_k^T \alpha(x_k) + V(x_{k+1}) - V(x_k)\} \\
&= -h^T(x_k) - \alpha^T(x_k)(J(x_k) + J^T(x_k)) + (f^T(x_k) - y_k^T g^T(x_k)) P g(x_k) \\
&= -h^T(x_k) - \alpha^T(x_k)(J(x_k) + J^T(x_k)) + f^T(x_k) P g(x_k) \\
&\quad - y_k^T g^T(x_k) P g(x_k)
\end{aligned}
\tag{3.14}
$$

where 0 is a vector of appropriate dimension with each entry equal to zero. Taking into account $h^T(x_k) = f^T(x_k) P g(x_k)$ (3.2), we obtain

$$
0 = -\alpha^T(x_k) J(x_k) - h^T(x_k) J(x_k) - \alpha^T(x_k) J^T(x_k) J(x_k)
$$

or

$$
\alpha^T(x_k)(J(x_k) + J^T(x_k) J(x_k)) = -h^T(x_k) J(x_k)
$$

and finally

$$
(J(x_k) + J^T(x_k) J(x_k)) \alpha(x_k) = -J(x_k) h(x_k).
\tag{3.15}
$$

From (3.15), the proposed control law immediately follows

$$
\alpha(x_k) = -(I_m + J(x_k))^{-1} h(x_k)
$$

as established in (3.1). In order to obtain the optimal value function ($\mathscr{J}^*$) for the cost functional (3.5), we solve (2.5) for $L(x_k, \alpha(x_k))$ and perform the addition over $[0, N]$, where $N \in 0 \cup \mathbb{N}$, then

$$
\sum_{k=0}^{N} L(x_k, \alpha(x_k)) = -V(x_N) + V(x_0) + \sum_{k=0}^{N} \mathscr{H}(x_k, \alpha(x_k)).
$$

From (2.6) we have $\mathscr{H}(x_k, \alpha(x_k)) = 0$ for the proposed optimal control $\alpha(x_k)$, and letting $N \to \infty$ and noting that $V(x_N) \to 0$ for any x_0, then $\mathscr{J}^* = V(x_0)$, which is called the optimal value function. Then $u_k = \alpha(x_k)$ is an inverse optimal control law which minimizes (3.5), with performance index $L(x_k, \alpha(x_k))$.   ∎

3.1.1 STABILIZATION OF A NONLINEAR SYSTEM

The applicability of the proposed control law is illustrated via simulation of an example. These simulations are performed using MATLAB®.[1]

We synthesize an inverse optimal control law for a discrete-time second order nonlinear system (unstable for $u_k = 0$) of the form (2.27) with

$$f(x_k) = \begin{bmatrix} 2.2\,sin(0.5x_{1,k}) + 0.1\,x_{2,k} \\ \\ 0.1\,x_{1,k}^2 + 1.8x_{2,k} \end{bmatrix} \tag{3.16}$$

and

$$g(x_k) = \begin{bmatrix} 0 \\ \\ 2 + 0.1\,cos(x_{2,k}) \end{bmatrix}. \tag{3.17}$$

According to (3.1), the inverse optimal control law is formulated as

$$\begin{aligned} u_k &= \alpha(x_k) \\ &= -\left(1 + \frac{1}{2}g^T(x_k)Pg(x_k)\right)^{-1} g^T(x_k)Pf(x_k) \end{aligned} \tag{3.18}$$

---

[1] Trademark of MathWorks, Inc., Natick, MA, USA.

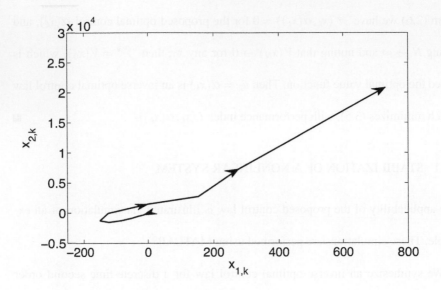

FIGURE 3.1   Unstable nonlinear system phase portrait.

for which we propose a positive definite matrix P as

$$P = \begin{bmatrix} 0.20 & 0.15 \\ 0.15 & 0.20 \end{bmatrix}.$$

In order to illustrate the applicability of the control scheme to stabilization, the phase portrait for this unstable open-loop $(u_k = 0)$ system with initial conditions $x_0 = [2, \, -2]^T$ is displayed in Figure 3.1.

Figure 3.2 presents the stabilized time response for x_k with initial conditions $x_0 = [2, \, -2]^T$; this figure also includes the applied inverse optimal control signal (3.18), which achieves asymptotic stability; the respective phase portrait is displayed in Figure 3.3.

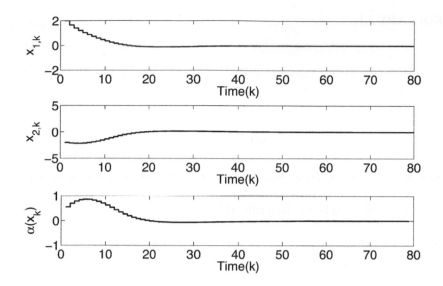

FIGURE 3.2   Stabilized nonlinear system time response.

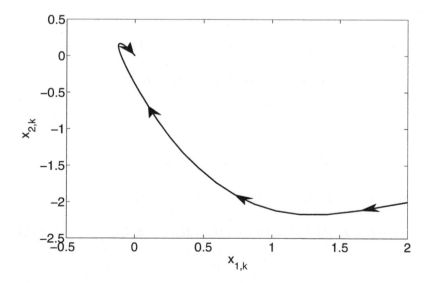

FIGURE 3.3   Stabilized nonlinear system phase portrait.

COROLLARY 3.1

Consider a class of nonlinear systems for which function $g(x_k)$ in (2.27) is a constant

matrix defined as

$$g(x_k) = B \tag{3.19}$$

of appropriate dimension. Then, the solution of inequality (3.4) for P reduces to the

solution of the following matrix inequality (MI):

$$\begin{bmatrix} P - 2PB\left(I_m + \tfrac{1}{2}B^T P B\right)^{-2} B^T P & 0 \\ \\ 0 & -P \end{bmatrix} < 0. \tag{3.20}$$

PROOF

From (3.1), $\alpha(x_k) = -(I_m + J(x_k))^{-1} h(x_k)$ with $h(x_k)$ and $J(x_k)$ as defined in (3.2)

and (3.3), respectively, the left-hand side of (3.4) reduces to

$$\left(f(x_k) + g(x_k)\,\alpha(x_k)\right)^T P \left(f(x_k) + g(x_k)\,\alpha(x_k)\right) - x_k^T P x_k$$

$$= f^T(x_k) P f(x_k) + 2 f^T(x_k) P g(x_k)\,\alpha(x_k) + \alpha^T(x_k) g^T(x_k) P g(x_k)\,\alpha(x_k) - x_k^T P x_k$$

$$= f^T(x_k) P f(x_k) - x_k^T P x_k - 2 h^T(x_k)\,(I_m + J(x_k))^{-1} h(x_k)$$

$$+ h^T(x_k)\,(I_m + J(x_k))^{-1} g^T(x_k) P g(x_k)\,(I_m + J(x_k))^{-1} h(x_k). \tag{3.21}$$

Considering $g^T(x_k)Pg(x_k) = 2(I_m+J(x_k))-2I_m$, (3.21) results in

$$f^T(x_k)Pf(x_k)-x_k^T Px_k-2h^T(x_k)(I_m+J(x_k))^{-1}h(x_k)$$

$$+h^T(x_k)(I_m+J(x_k))^{-1}[2(I_m+J(x_k))-2I_m](I_m+J(x_k))^{-1}h(x_k)$$

$$= f^T(x_k)Pf(x_k)-x_k^T Px_k-2h^T(x_k)(I_m+J(x_k))^{-2}h(x_k)$$

$$= f^T(x_k)Pf(x_k)-x_k^T Px_k-2f^T(x_k)Pg(x_k)\left(I_m+\frac{1}{2}g^T(x_k)Pg(x_k)\right)^{-2}$$

$$\times g^T(x_k)Pf(x_k). \tag{3.22}$$

Taking into account condition (3.19), then (3.4) is satisfied by determining $P=P^T>0$

for (3.22) such that

$$f^T(x_k)Pf(x_k)-x_k^T Px_k-2f^T(x_k)Pg(x_k)\left(I_m+\frac{1}{2}g^T(x_k)Pg(x_k)\right)^{-2}g^T(x_k)Pf(x_k)$$

$$= f^T(x_k)Pf(x_k)-x_k^T Px_k-2f^T(x_k)PB\left(I_m+\frac{1}{2}B^T PB\right)^{-2}B^T Pf(x_k)$$

$$= f^T(x_k)\left[P-2PB\left(I_m+\frac{1}{2}B^T PB\right)^{-2}B^T P\right]f(x_k)-x_k^T Px_k$$

$$= \begin{bmatrix} f(x_k) & x_k \end{bmatrix}^T \begin{bmatrix} P-2PB\left(I_m+\frac{1}{2}B^T PB\right)^{-2}B^T P & 0 \\ & \\ 0 & -P \end{bmatrix} \begin{bmatrix} f(x_k) \\ x_k \end{bmatrix}$$

$$< 0. \tag{3.23}$$

∎

REMARK 3.1 Different nonlinear systems satisfy condition (3.19) such as electrical

machines (induction motors [82, 23], DC motors with controlled excitation flux [20],

synchronous generators [82]), mechanical systems [82] (spacecraft systems, ball and bean systems, robots with flexible joints, etc.), models obtained from neural iden-

tification schemes [116] for which condition (3.19) can be imposed, among others.

■

3.2 TRAJECTORY TRACKING

For many applications, such as aerospace, electric machinery, robotics, among others,

it is important for control purposes to track a desired trajectory. In order to achieve

such a control objective, we extend our approach as follows. We modify the storage

function, used as CLF, such that this new function has a global minimum along the

desired trajectory $x_{\delta,k}$. Then, we redefine the CLF as

$$V(x_k, x_{\delta,k}) = \frac{1}{2}(x_k - x_{\delta,k})^T K^T P K (x_k - x_{\delta,k}) \tag{3.24}$$

where $x_{\delta,k}$ is the desired trajectory and K is an additional gain matrix introduced to

modify the convergence rate of the tracking error.

Theorem 3.2

Assume an affine discrete-time nonlinear system (2.27) and an output defined as

$$y_k = h(x_k, x_{\delta,k+1}) + J(x_k) u_k \tag{3.25}$$

which is zero-state detectable. Consider the passivity condition

$$V(x_{k+1}, x_{\delta,k+1}) - V(x_k, x_{\delta,k}) \le y_k^T u_k \tag{3.26}$$

with (3.24) as a CLF candidate.

If there exists $\overline{P} = \overline{P}^T > 0$ such that the following condition holds

$$f^T(x_k)\overline{P}f(x_k) + x_{\delta,k+1}^T \overline{P}x_{\delta,k+1} - 2f^T(x_k)\overline{P}x_{\delta,k+1} - (x_k - x_{\delta,k})^T \overline{P}(x_k - x_{\delta,k}) \leq 0$$

$$(3.27)$$

where $\overline{P} = K^T PK$, then the solution of system (2.27) with output (3.25) is passive, and global asymptotic stability along the desired trajectory $x_{\delta,k}$ is ensured by the output feedback

$$
\begin{aligned}
u_k &= -y_k \\
&= -(I_m + J(x_k))^{-1} h(x_k, x_{\delta,k+1})
\end{aligned}
$$
$$(3.28)$$

with

$$h(x_k, x_{\delta,k+1}) = g^T(x_k)\overline{P}(f(x_k) - x_{\delta,k+1})$$

and

$$J(x_k) = \frac{1}{2}g^T(x_k)\overline{P}g(x_k).$$

Moreover, with (3.24) as a CLF, control law (3.28) is inverse optimal in the sense that it minimizes the cost functional

$$\mathscr{J} = \sum_{k=0}^{\infty} L(x_k, x_{\delta,k}, u_k)$$

$$(3.29)$$

where $L(x_k, x_{\delta,k}, u_k)$ is a nonnegative function. ■

PROOF

Let (3.24) be a CLF candidate. System (2.27) with output (3.25) must be rendered passive, such that the inequality (3.26) is fulfilled. Thus, from (3.26), and considering one step ahead for $x_{\delta,k}$, we have

$$\frac{(x_{k+1}-x_{\delta,k+1})^T K^T P K (x_{k+1}-x_{\delta,k+1})}{2} - \frac{(x_k-x_{\delta,k})^T K^T P K (x_k-x_{\delta,k})}{2}$$

$$\leq h^T(x_k,x_{\delta,k})u_k + u_k^T J^T(x_k)u_k. \qquad (3.30)$$

Substituting (2.27) in (3.30), we obtain

$$\frac{(f(x_k)+g(x_k)u_k-x_{\delta,k+1})^T \overline{P}(f(x_k)+g(x_k)u_k-x_{\delta,k+1})}{2}$$

$$- \frac{(x_k-x_{\delta,k})^T \overline{P}(x_k-x_{\delta,k})}{2} \leq h^T(x_k,x_{\delta,k})u_k + u_k^T J^T(x_k)u_k. \qquad (3.31)$$

Then, (3.31) becomes

$$f^T(x_k)\overline{P}f(x_k) + x_{\delta,k+1}^T \overline{P}x_{\delta,k+1} - 2f^T(x_k)\overline{P}x_{\delta,k+1} - (x_k-x_{\delta,k})^T \overline{P}(x_k-x_{\delta,k})$$

$$+2\left(f^T(x_k)\overline{P}g(x_k) - x_{\delta,k+1}^T \overline{P}g(x_k)\right)u_k + u_k^T g^T(x_k)\overline{P}g(x_k)u_k \qquad (3.32)$$

$$\leq 2h^T(x_k,x_{\delta,k})u_k + 2u_k^T J^T(x_k)u_k.$$

Passivity condition (3.32) is rewritten in accordance with related terms for both equation sides as follows:

- From the first term of (3.32), we obtain

$$f^T(x_k)\overline{P}f(x_k) + x_{\delta,k+1}^T \overline{P}x_{\delta,k+1} - 2f^T(x_k)\overline{P}x_{\delta,k+1}$$

$$-(x_k-x_{\delta,k})^T \overline{P}(x_k-x_{\delta,k}) \leq 0. \qquad (3.33)$$

- $2\left(f^T(x_k)\overline{P}g(x_k) - x_{\delta,k+1}^T \overline{P}g(x_k)\right)u_k = 2h^T(x_k,x_{\delta,k})u_k$, then we define $h(x_k,x_{\delta,k+1})$ as

$$h(x_k,x_{\delta,k+1}) = g^T(x_k)\overline{P}\left(f(x_k) - x_{\delta,k+1}\right). \qquad (3.34)$$

- $u^T g^T(x_k)\overline{P}g(x_k)u_k = 2u_k^T J^T(x_k)u_k$, then we define $J(x_k)$ as

$$J(x_k) = \frac{1}{2}g^T(x_k)\overline{P}g(x_k). \qquad (3.35)$$

From (3.33)–(3.35) we deduce that, if there exists a $\overline{P}$ such that (3.33) is satisfied, then system (2.27) with output (3.25) is passive, with $h(x_k,x_{\delta,k+1})$ and $J(x_k)$ as defined in (3.34) and (3.35), respectively. To guarantee asymptotic trajectory tracking, we select $u_k = -y_k$ and then condition (3.26) redefined as $V(x_{k+1},x_{\delta,k+1}) - V(x_k,x_{\delta,k}) \leq -y_k^T y_k < 0 \ \forall y_k \neq 0$ is fulfilled.

Defining $e_k = x_k - x_{\delta,k}$, (3.24) can be rewritten as

$$V(e_k) = \frac{1}{2}e_k^T \overline{P}e_k \qquad (3.36)$$

which is a radially unbounded function w.r.t. e_k. Hence, the solution of the closed-loop system (2.27) with (3.28) as input is globally asymptotically stable along the desired trajectory $x_{\delta,k}$.

The minimization of the cost functional is established similarly as in Theorem 3.1, and hence it is omitted. ∎

3.2.1 EXAMPLE: TRAJECTORY TRACKING OF A NONLINEAR SYSTEM

The applicability of the proposed trajectory tracking control scheme is illustrated via simulation.

In accordance with Theorem 3.2, the control law for trajectory tracking is given in

(3.28) as $u_k = -y_k$, in which $y_k = h(x_k, x_{\delta,k+1}) + J(x_k) u_k$, where

$$h(x_k, x_{\delta,k+1}) = g^T(x_k)\overline{P}(f(x_k) - x_{\delta,k+1})$$

and

$$J(x_k) = \frac{1}{2}g^T(x_k)\overline{P}g(x_k)$$

with $f(x_k)$ and $g(x_k)$ as defined in (3.16) and (3.17), respectively. Hence, we adjust

gain matrix $\overline{P} = K^T P K$ for (3.28) in order to achieve trajectory tracking for $x_k =$

$[x_{1,k}\ x_{2,k}]^T$. The reference for $x_{2,k}$ is

$$x_{2\delta,k} = 1.5\sin(0.12k)\ rad.$$

and reference $x_{1\delta,k}$ is defined accordingly.

Figure 3.4 presents trajectory tracking results, where the solid line $(x_{\delta,k})$ is the

reference signal and the dashed line is the evolution of x_k. The control signal is also

displayed. Matrices P and K are selected as

$$P = \begin{bmatrix} 0.00340 & 0.00272 \\ 0.00272 & 0.00240 \end{bmatrix};\ K = \begin{bmatrix} 0.10 & 0 \\ 0 & 12.0 \end{bmatrix}.$$

3.2.2   APPLICATION TO A PLANAR ROBOT

In this section, we apply Section 3.2 results to synthesize position trajectory tracking

control for a two Degrees of Freedom (DOF) planar rigid robot manipulator.

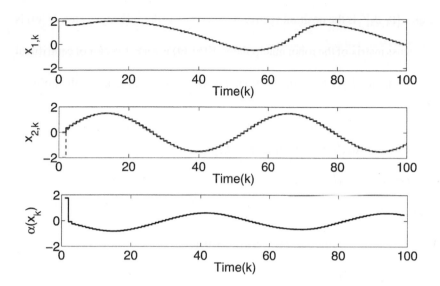

FIGURE 3.4 (SEE COLOR INSERT)   Nonlinear system tracking performance.

3.2.2.1   Robot Model

We assume that the robot dynamics [25] is given by

$$\ddot{\Theta} = M^{-1}(\Theta)\left[\tau - V(\Theta,\dot{\Theta}) - F(\Theta,\dot{\Theta})\right] \tag{3.37}$$

with

$$M(\Theta) = \begin{bmatrix} D_{11}(\Theta) & D_{12}(\Theta) \\ D_{12}(\Theta) & D_{22}(\Theta) \end{bmatrix} \tag{3.38}$$

$$V(\Theta,\dot{\Theta}) = \begin{bmatrix} V_1(\Theta,\dot{\Theta}) \\ V_2(\Theta,\dot{\Theta}) \end{bmatrix}$$

$$F(\Theta,\dot{\Theta}) = \begin{bmatrix} F_1(\Theta,\dot{\Theta}) \\ F_2(\Theta,\dot{\Theta}) \end{bmatrix}$$

where $\Theta = [\theta_1 \ \theta_2]^T$ is the position vector, $\dot{\Theta} = [\dot{\theta}_1 \ \dot{\theta}_2]^T$ is the velocity vector, $M(\Theta)$ is

the $n \times n$ mass matrix of the robot manipulator, $V(\Theta, \dot{\Theta})$ is a $n \times 1$ vector of centrifugal

and Coriolis terms, $F(\Theta, \dot{\Theta})$ is the friction, τ is the torque vector, and with entries in

(3.37) as

$$D_{11}(\Theta) = m_1 l_{c1}^2 + m_2(l_1^2 + l_{c2}^2 + 2l_1 l_{c2} c_2) + I_{zz_1} + I_{zz_2}$$

$$D_{12}(\Theta) = m_2 l_{c2}^2 + m_2 l_1 l_{c2} c_2 + I_{zz_2}$$

$$D_{22}(\Theta) = m_2 l_{c2}^2 + I_{zz_2}$$

$$V_1(\Theta, \dot{\Theta}) = -m_2 l_1 l_{c2} s_2 (\dot{\theta}_1 + \dot{\theta}_2) \dot{\theta}_2 - m_2 l_1 l_{c2} \dot{\theta}_1 \dot{\theta}_2 s_2$$

$$V_2(\Theta, \dot{\Theta}) = m_2 l_1 l_{c2} s_2 (\dot{\theta}_1)^2$$

$$F_1(\Theta, \dot{\Theta}) = \mu_1 \dot{\theta}_1$$

$$F_2(\Theta, \dot{\Theta}) = \mu_2 \dot{\theta}_2$$

where $s_2 = sin(\theta_2)$, $c_2 = cos(\theta_2)$. The described planar robot is depicted in Figure

3.5 and the respective free-body diagram in XY-coordinates is shown in Figure 3.6.

Let us define new variables as $x_1 = \theta_1$, $x_2 = \theta_2$, $x_3 = \dot{\theta}_1$, $x_4 = \dot{\theta}_2$, and after dis-

cretizing by Euler approximation,[2] the discrete-time planar robot model is written

as

$$x_{1,k+1} = x_{1,k} + x_{3,k} T$$

$$x_{2,k+1} = x_{2,k} + x_{4,k} T$$

---

[2]For the ordinary differential equation $\frac{dz}{dt} = f(z)$, the Euler discretization is defined as $\frac{z_{k+1} - z_k}{T} = f(z_k)$, such that $z_{k+1} = z_k + T f(z_k)$, where T is the sampling time [67, 74].

FIGURE 3.5   Two DOF planar robot.

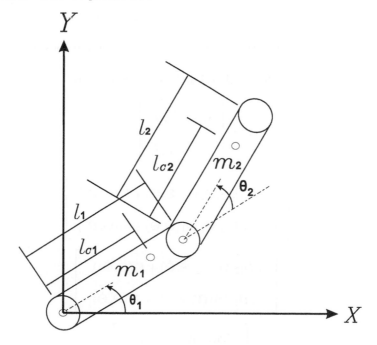

FIGURE 3.6   Free-body diagram of the two DOF planar robot.

$$
\begin{aligned}
x_{3,k+1} &= x_{3,k} + \left(\frac{-D_{22}(\Theta)\left(V_1(\Theta,\dot{\Theta})+F_1(\Theta,\dot{\Theta})\right)}{D_{11}(\Theta)D_{22}(\Theta)-D_{12}(\Theta)^2} \right. \\
&\quad + \left. \frac{D_{12}(\Theta)\left(V_2(\Theta,\dot{\Theta})+F_2(\Theta,\dot{\Theta})\right)+D_{22}(\Theta)u_{1,k}-D_{12}(\Theta)u_{2,k}}{D_{11}(\Theta)D_{22}(\Theta)-D_{12}(\Theta)^2} \right) T \\
x_{4,k+1} &= x_{4,k} + \left(\frac{D_{12}(\Theta)\left(V_1(\Theta,\dot{\Theta})+F_1(\Theta,\dot{\Theta})\right)}{D_{11}(\Theta)D_{22}(\Theta)-D_{12}(\Theta)^2} \right. \\
&\quad + \left. \frac{-D_{11}(\Theta)\left(V_2(\Theta,\dot{\Theta})+F_2(\Theta,\dot{\Theta})\right)-D_{12}(\Theta)u_{1,k}+D_{11}(\Theta)u_{2,k}}{D_{11}(\Theta)D_{22}(\Theta)-D_{12}(\Theta)^2} \right) T.
\end{aligned}
\tag{3.39}
$$

where T is the sampling time, and $u_{1,k}$ and $u_{2,k}$ are the applied torques.

3.2.2.2 Robot as an Affine System

In order to make the controller synthesis easier, we rewrite (3.39) in a block structure form as

$$
\begin{aligned}
x_{12,k+1} &= f_1(x_k) \\
x_{34,k+1} &= f_2(x_k)+g(x_k)u(x_k), \qquad x_0 = x(0)
\end{aligned}
\tag{3.40}
$$

where $x_k = [x_{12,k}^T \ x_{34,k}^T]^T$, being $x_{12,k} = [x_{1,k} \ x_{2,k}]^T$ the position variables, and $x_{34,k} = [x_{3,k} \ x_{4,k}]^T$ the velocity variables for link 1 and link 2, respectively,

$$
f_1(x_k) = \begin{bmatrix} x_{1,k}+x_{3,k}T \\ x_{2,k}+x_{4,k}T \end{bmatrix};
$$

$$
f_2(x_k) = \begin{bmatrix} x_{3,k}-cD_{22}(\Theta)\left(V_1(\Theta,\dot{\Theta})+F_1(\Theta,\dot{\Theta})\right) \\ x_{4,k}+cD_{12}(\Theta)\left(V_1(\Theta,\dot{\Theta})+F_1(\Theta,\dot{\Theta})\right) \end{bmatrix}
$$
$$
+ \begin{bmatrix} cD_{12}(\Theta)\left(V_2(\Theta,\dot{\Theta})+F_2(\Theta,\dot{\Theta})\right) \\ -cD_{11}(\Theta)\left(V_2(\Theta,\dot{\Theta})+F_2(\Theta,\dot{\Theta})\right) \end{bmatrix};
$$

$$
g(x_k) = \begin{bmatrix} D_{22}(\Theta) & -D_{12}(\Theta) \\ -D_{12}(\Theta) & D_{11}(\Theta) \end{bmatrix}
$$

with $c = T/(D_{11}(\Theta)D_{22}(\Theta) - D_{12}(\Theta)^2) \neq 0$ being the denominator of the robot mass

inverse matrix $M^{-1}(\Theta)$. Note that any mass matrix for manipulators is symmetric and

positive definite, and therefore it is always invertible [25].

3.2.2.3 Control Synthesis

For trajectory tracking, we propose the CLF as

$$V(x_k, x_{\delta,k}) = \frac{1}{2}(x_k - x_{\delta,k})^T K^T P K (x_k - x_{\delta,k})$$

where $x_{\delta,k}$ are the reference trajectories, K is an additional gain to modify the con-

vergence rate, and matrix P is synthesized to achieve passivity according to Section

3.2, which can be written, respectively, with a block structure as:

$$K = \begin{bmatrix} K_1 & 0 \\ 0 & K_2 \end{bmatrix}$$

and

$$P = \begin{bmatrix} P_{11} & P_{12} \\ P_{21} & P_{22} \end{bmatrix}.$$

Hence, from passivity condition (3.26) for system (3.40), and according to (3.34)–

(3.35), the output is established as

$$y_k = h(x_k, x_{\delta,k+1}) + J(x_k)u_k$$

where

$$h(x_k, x_{\delta,k+1}) = g^T(x_k)\left(P_{22} f_2(x_k) - K_1^T P_{12} x_{12\delta,k+1} + K_1^T P_{12} f_1(x_k)\right)$$

and

$$J(x_k) = \frac{1}{2} g^T(x_k) P_{22} g(x_k).$$

Global asymptotic convergence to state reference trajectory is guaranteed with (3.28),

if we can find a positive definite matrix $\overline{P}$ satisfying (3.27).

3.2.2.4 Simulation Results

The parameters of the planar robot used for simulation are given in Table 3.1. The

reference signals are

$$x_{1\delta,k} = 2.0\sin(1.0kT) \; rad$$

$$x_{2\delta,k} = 1.5\sin(1.2kT) \; rad.$$

References $x_{3\delta,k}$ and $x_{4\delta,k}$, are defined accordingly.

These signals are selected to illustrate the ability of the proposed algorithm to track

nonlinear trajectories.

Equation (3.27) is satisfied with

$$P_{11} = 100 * \begin{bmatrix} 4 & 3 \\ 3 & 4 \end{bmatrix} ; \; P_{12} = 100 * \begin{bmatrix} 2 & 1 \\ 3 & 2 \end{bmatrix}$$

$$P_{21} = P_{12}^T; \; P_{22} = P_{11}^T; \; K_1 = \begin{bmatrix} 170 & 0 \\ 0 & 110 \end{bmatrix}$$

and K_2 is chosen as the 2×2 identity matrix.

Tracking performance for both link 1 and link 2 positions are shown in Figure 3.7,

where $x_{12\delta}$ (*solid line*) are the reference signals and x_{12} (*dashed line*) are the link

TABLE 3.1

Model parameters.

PARAMETER	VALUE	DESCRIPTION
l_1	0.3 m	Length of the link 1
l_{c1}	0.2 m	Mean length of the link 1
l_2	0.25 m	Length of the link 2
l_{c2}	0.1 m	Mean length of the link 2
m_1	1 Kg	Mass of the link 1
m_2	0.3 Kg	Mass of the link 2
I_{zz1}	0.05 $Kg - m^2$	Moment of inertia 1
I_{zz2}	0.004 $Kg - m^2$	Moment of inertia 2
μ_1	0.005 Kg/s	Friction coefficient 1
μ_2	0.0047 Kg/s	Friction coefficient 2

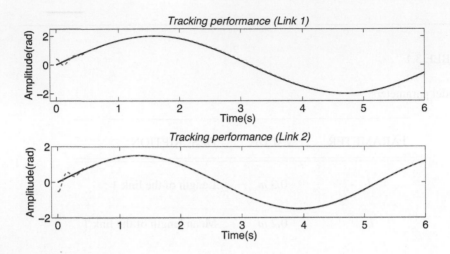

FIGURE 3.7 **(SEE COLOR INSERT)**   Robot tracking performance.

positions. The initial conditions for the robot manipulator are $x_{1,k} = 0.4 \ rad; x_{2,k} = -0.5 \ rad; x_{3,k} = 0 \ rad/s; x_{4,k} = 0$; and $T = 0.001$.

Control signals u_1 and u_2 are displayed in Figure 3.8.

3.3   PASSIVITY-BASED INVERSE OPTIMAL CONTROL FOR A CLASS

OF NONLINEAR POSITIVE SYSTEMS

Diverse physical systems commonly involve variables such as mass, pressure, population levels, energy, humidity, etc., which are always positive. Such systems are called positive systems, i.e., systems whose states and outputs are always nonnegative provided that the initial conditions and the input are nonnegative [15, 31, 45, 78]. Mainly, the models representing biological processes are positive systems; perhaps the most natural examples of positive systems are the models obtained from mass balances, like the compartmental ones. This type of model is used to describe transportation,

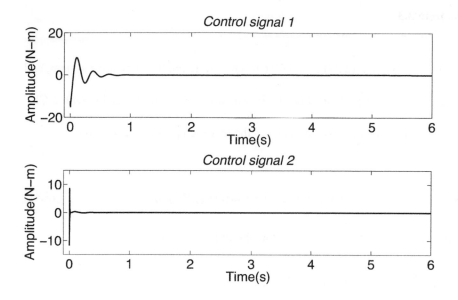

FIGURE 3.8   Control signals u_1 and u_2 time evolution.

and accumulation and drainage processes of elements and compounds like hormones, glucose, insulin, metals, etc. [31].

Positive systems have been analyzed recently; stability is one of the most important properties and many authors have studied this feature for linear positive systems [31, 45, 62, 112]. However, there are not many publications for the case of trajectory tracking of nonlinear positive systems.

Considering this class of systems and existing results, the goal of this section is to design a feedback controller which stabilizes a nonlinear positive system along a desired trajectory. The controller, based on the inverse optimal control approach presented in [93], is extended to a class of positive systems.

Theorem 3.3

Assume that system (2.27) is an affine discrete-time nonlinear positive system, and consider the output as defined in (3.25), which is zero-state detectable with a CLF candidate defined by (3.24), and satisfies the modified passivity condition

$$V(x_{k+1}, x_{\delta,k+1}) - V(x_k, x_{\delta,k}) \leq y_k^T u_k. \tag{3.41}$$

If there exists a $\overline{P} = \overline{P}^T > 0$ such that the following condition holds

$$f^T(x_k)\overline{P}f(x_k) + x_{\delta,k+1}^T \overline{P}x_{\delta,k+1} - f^T(x_k)\overline{P}x_{\delta,k+1}$$

$$-x_{\delta,k+1}^T \overline{P}f(x_k) - (x_k - x_{\delta,k})^T \overline{P}(x_k - x_{\delta,k}) \leq 0 \tag{3.42}$$

where $\overline{P} = K_1^T P_1 K_1$ is a definite positive matrix, then system (2.27) with output (3.25) is globally asymptotically stabilized along the desired trajectory $x_{\delta,k}$ by the output feedback

$$u_k = \left| (I_m + J(x_k))^{-1} h(x_k, x_{\delta,k+1}) \right| \tag{3.43}$$

with

$$h(\bullet) = \begin{cases} g^T(x_k)\overline{P}(x_{\delta,k+1} - f(x_k)) & \text{for} \quad f(x_k) \succeq x_{\delta,k+1} \\ g^T(x_k)\overline{P}(f(x_k) - x_{\delta,k+1}) & \text{for} \quad f(x_k) \preceq x_{\delta,k+1} \end{cases} \tag{3.44}$$

and

$$J(x_k) = \frac{1}{2}g^T(x_k)\overline{P}g(x_k). \tag{3.45}$$

Moreover, with (3.24) as a CLF, this control law is inverse optimal in the sense that it minimizes the functional cost as given in (3.29) [94]. ∎

PROOF

Case 1. Consider the case for which $h(x_k, x_{\delta,k+1}) = g^T(x_k)\overline{P}(x_{\delta,k+1} - f(x_k))$. Let (3.24) be a CLF candidate. System (2.27) with output (3.25) must be rendered passive, such that the inequality (3.41) is fulfilled. Then, from (3.41), and considering one step ahead for $x_{\delta,k}$, we have

$$
\begin{aligned}
& \frac{(x_{\delta,k+1} - x_{k+1})^T K_1^T P_1 K_1 (x_{\delta,k+1} - x_{k+1})}{2} \\
& - \frac{(x_{\delta,k} - x_k)^T K_1^T P_1 K_1 (x_{\delta,k} - x_k)}{2} \\
& \leq \; h^T(x_k, x_{\delta,k}) u_k + u_k^T J^T(x_k) u_k.
\end{aligned}
\tag{3.46}
$$

Defining $\overline{P} = K_1^T P_1 K_1$ and substituting (2.27) in (3.46), we have

$$
\begin{aligned}
& \frac{(x_{\delta,k+1} - f(x_k) - g(x_k)u_k)^T \overline{P}(x_{\delta,k+1} - f(x_k) - g(x_k)u_k)}{2} \\
& - \frac{(x_{\delta,k} - x_k)^T \overline{P}(x_{\delta,k} - x_k)}{2} \\
& \leq \; h^T(x_k, x_{\delta,k}) u_k + u_k^T J^T(x_k) u_k.
\end{aligned}
\tag{3.47}
$$

Hence, (3.47) becomes

$$
\begin{aligned}
& f^T(x_k)\overline{P}f(x_k) + x_{\delta,k+1}^T \overline{P}x_{\delta,k+1} - f^T(x_k)\overline{P}x_{\delta,k+1} \\
& - x_{\delta,k+1}^T \overline{P}f(x_k) - (x_{\delta,k} - x_k)^T \overline{P}(x_{\delta,k} - x_k) \\
& + (2f^T(x_k)\overline{P}g(x_k) - 2x_{\delta,k+1}^T \overline{P}g(x_k))u_k + u_k^T g^T(x_k)\overline{P}g(x_k)u_k \\
& \leq \; 2h^T(x_k, x_{\delta,k+1}) u_k + 2u_k^T J^T(x_k) u_k.
\end{aligned}
\tag{3.48}
$$

From (3.48), passivity is achieved if

1) from the first term of (3.48), we can obtain $\overline{P} > 0$ such that

$$f^T(x_k)\overline{P}f(x_k) + x_{\delta,k+1}^T\overline{P}x_{\delta,k+1} - f^T(x_k)\overline{P}x_{\delta,k+1}$$

$$-x_{\delta,k+1}^T\overline{P}f(x_k) - (x_k - x_{\delta,k})^T\overline{P}(x_k - x_{\delta,k}) \leq 0; \qquad (3.49)$$

2) with $(2f^T(x_k)\overline{P}g(x_k) - 2x_{\delta,k+1}^T\overline{P}g(x_k))u_k = 2h^T(x_k, x_{\delta,k+1})u_k$, then

$$h(x_k, x_{\delta,k+1}) = g^T(x_k)\overline{P}(f(x_k) - x_{\delta,k+1}); \qquad (3.50)$$

3) and $u_k^T g^T(x_k)\overline{P}g(x_k)u_k = 2u_k^T J^T(x_k)u_k$, then

$$J(x_k) = \frac{1}{2}g^T(x_k)\overline{P}g(x_k). \qquad (3.51)$$

Let system (2.27) with output (3.25) fulfill the zero-state detectability property; if $\overline{P}$ satisfies (3.49), then from 1), 2), and 3) we deduce that system (2.27) with output (2.28) is passive.

To guarantee asymptotic trajectory tracking, we select $u_k = -y_k$ and then $V(x_{k+1}, x_{\delta,k+1}) - V(x_k, x_{\delta,k}) \leq -y_k^T y_k \leq 0$, which satisfies the Lyapunov forward difference of V.

In order to establish the inverse optimality, (2.5) is minimized w.r.t. u_k, with

$$L(x_k, u_k) = l(x_k) + u_k^T u_k = l(x_k) - y_k^T u_k$$

where $l(x_k) = -\left(f^T(x_k)P_1f(x_k) - x_k^T Px_k\right) \geq 0$; thus, we have

$$
\begin{aligned}
0 &= \min_{u_k}\{L(x_k, u_k) + V(x_{k+1}) - V(x_k)\} \\
&= \min_{u_k}\{l(x_k) - y_k^T u_k + V(x_{k+1}) - V(x_k)\} \\
&= -h^T(x_k, x_{\delta,k+1}) - u_k^T\left(J(x_k) + J^T(x_k)\right) + \left(f^T(x_k) - y_k^T g^T(x_k)\right)P_1 g(x_k) \\
&= -h^T(x_k, x_{\delta,k+1}) - u_k^T\left(J(x_k) + J^T(x_k)\right) + f^T(x_k)P_1 g(x_k) \\
&\quad - y_k^T g^T(x_k)P_1 g(x_k).
\end{aligned}
$$

Considering $h^T(x_k, x_{\delta,k+1}) = f^T(x_k) P_1 g(x_k)$ and $J(x_k) = J^T(x_k)$, then

$$0 \; = \; -u_k^T J(x_k) - h^T(x_k, x_{\delta,k+1}) J(x_k) - u_k^T J^T(x_k) J(x_k)$$

$$u_k^T \left(J(x_k) + J^T(x_k) J(x_k) \right) \; = \; -h^T(x_k, x_{\delta,k+1}) J(x_k)$$

$$\left(J(x_k) + J^2(x_k) \right) u_k \; = \; -J(x_k) h(x_k, x_{\delta,k+1})$$

and solving for u_k, the proposed inverse optimal control law is obtained

$$u_k = -\left(I_m + J(x_k) \right)^{-1} h\left(x_k, x_{\delta,k+1} \right) \tag{3.52}$$

with $h(x_k, x_{\delta,k+1}) = g^T(x_k) \overline{P}(x_{\delta,k+1} - f(x_k))$.

Now, we solve (2.5) for $L(x_k, u_k)$ and sum over $[0, N]$, where $N \in 0 \cup \mathbb{N}$ yields

$$\sum_{k=0}^{N} L(x_k, u_k) = -V(x_N) + V(x_0) + \sum_{k=0}^{N} \mathcal{H}(x_k, u_k).$$

Letting $N \to \infty$ and noting that $V(x_N) \longrightarrow 0$ for all x_0, and $\mathcal{H}(x_k, u_k) = 0$ for the inverse optimal control u_k, then $C(x_0, u_k) = V(x_0)$, which is named the optimal value function. Finally, if $V(x_k)$ is a radially unbounded function, i.e., $V(x_k) \longrightarrow \infty$ as $\|x_k\| \longrightarrow \infty$, then the solution $x_k = 0$ of the closed-loop system (2.27) with (3.43) is globally asymptotically stable.

Case 2. Consider the case for which $h(x_k, x_{\delta,k+1}) = g^T(x_k) \overline{P}(f(x_k) - x_{\delta,k+1})$ [93]. It can be derived as explained in [93]. Let (3.24) be a CLF candidate. System (2.27) with output (3.25) must be rendered passive, such that the inequality (3.41) is fulfilled. Then, from (3.41), and considering one step ahead for $x_{\delta,k}$, we have

$$\begin{aligned}
&\frac{(x_{k+1} - x_{\delta,k+1})^T K_1^T P_1 K_1 (x_{k+1} - x_{\delta,k+1})}{2} \\
&-\frac{(x_k - x_{\delta,k})^T K_1^T P_1 K_1 (x_k - x_{\delta,k})}{2} \\
&\leq \; h^T(x_k, x_{\delta,k}) u_k + u_k^T J^T(x_k) u_k.
\end{aligned} \tag{3.53}$$

Defining $\overline{P} = K_1^T P_1 K_1$ and substituting (2.27) in (3.53), then

$$\frac{(f + g u_k - x_{\delta,k+1})^T \overline{P}(f + g u_k - x_{\delta,k+1})}{2}$$

$$-\frac{(x_k - x_{\delta,k})^T \overline{P}(x_k - x_{\delta,k})}{2} \tag{3.54}$$

$$\leq \quad h^T(x_k, x_{\delta,k})u_k + u_k^T J^T(x_k)u_k.$$

Hence, (3.54) becomes

$$f^T(x_k)\overline{P}f(x_k) + x_{\delta,k+1}^T \overline{P}x_{\delta,k+1} - f^T(x_k)\overline{P}x_{\delta,k+1}$$

$$-x_{\delta,k+1}^T \overline{P}f(x_k) - (x_k - x_{\delta,k})^T \overline{P}(x_k - x_{\delta,k}) \tag{3.55}$$

$$+(2f^T(x_k)\overline{P}g(x_k) - 2x_{\delta,k+1}^T \overline{P}g(x_k))u_k + u_k^T g^T(x_k)\overline{P}g(x_k)u_k$$

$$\leq \quad 2h^T(x_k, x_{\delta,k})u_k + 2u_k^T J^T(x_k)u_k.$$

From (3.55), passivity is achieved if

1) from the first term of (3.55), we can obtain $\overline{P} > 0$ such that

$$f^T(x_k)\overline{P}(x_k)f + x_{\delta,k+1}^T \overline{P}x_{\delta,k+1} - f^T(x_k)\overline{P}x_{\delta,k+1}$$

$$-x_{\delta,k+1}^T \overline{P}f(x_k) - (x_k - x_{\delta,k})^T \overline{P}(x_k - x_{\delta,k}) \leq 0; \tag{3.56}$$

2) with $(2f^T(x_k)\overline{P}g(x_k) - 2x_{\delta,k+1}^T \overline{P}g(x_k))u_k = 2h^T(x_k, x_{\delta,k+1})u_k$, thus

$$h(x_k, x_{\delta,k+1}) = g^T(x_k)\overline{P}(f(x_k) - x_{\delta,k+1}); \tag{3.57}$$

3) and $u^T g^T(x_k)\overline{P}g(x_k)u_k = 2u_k^T J^T(x_k)u_k$, thus

$$J(x_k) = \frac{1}{2}g^T(x_k)\overline{P}g(x_k). \tag{3.58}$$

Let (2.27) with output (3.25) fulfill the zero-state detectability property; if $\overline{P}$ satisfies (3.56), then from 1), 2), and 3) we deduce that system (2.27) with output (2.28) is passive. To guarantee asymptotic trajectory tracking, we select $u_k = -y_k$ and then

$V(x_{k+1}, x_{\delta,k+1}) - V(x_k, x_{\delta,k}) \leq -y_k^T y_k \leq 0$, which satisfies the Lyapunov forward

difference of V. Besides, the inverse optimality is established similarly to Case 1.

Therefore the proposed inverse optimal control law is given as

$$u_k = -\left(I_m + J(x_k)\right)^{-1} h\left(x_k, x_{\delta,k+1}\right) \tag{3.59}$$

with $h(x_k, x_{\delta,k+1}) = g^T(x_k) \overline{P}(f(x_k) - x_{\delta,k+1})$.

Finally, combining (3.52) and (3.59) the control law is given as

$$u_k = \left| -\left(I_m + J(x_k)\right)^{-1} h(x_k, x_{\delta,k+1}) \right| \tag{3.60}$$

which ensures that $h(\cdot)$ satisfies (3.44), where $|\cdot|$ denotes absolute value. ∎

Applicability of the previous results are illustrated for glucose control in Chapter

7.

3.4 CONCLUSIONS

This chapter has presented a novel discrete-time inverse optimal control, which

achieves stabilization and trajectory tracking for a nonlinear systems and is inverse

optimal in the sense that it, a posteriori, minimizes a cost functional. The controller

synthesis is based on the selection of a CLF and a passifying law to render the system

passive. The applicability of the proposed methods is illustrated by means of one

example and the applicability to a planar robot. The example is an unstable nonlinear

system in which stabilization and trajectory tracking is achieved, and for the planar

robot, trajectory tracking is accomplished. The inverse optimal control via passivity

is also extended to a class of positive nonlinear systems.

4 Inverse Optimal Control:

A CLF Approach, Part I

In this chapter, we establish inverse optimal control and its solution by proposing a quadratic CLF in Section 4.1. The CLF depends on a fixed parameter in order to satisfy stability and optimality conditions. In Section 4.2, a robust inverse optimal control is proposed for a disturbed nonlinear system. The inverse optimal control is extended to achieve trajectory tracking in Section 4.3. Additionally, in Section 4.4 the inverse optimal control technique for positive systems based on the CLF approach is proposed. Simulation results illustrate the applicability of the proposed control schemes.

4.1 INVERSE OPTIMAL CONTROL VIA CLF

Due to favorable stability margins of optimal control systems, we synthesize a stabilizing feedback control law, which will be optimal with respect to a cost functional. At the same time, we want to avoid the difficult task of solving the HJB partial differential equation. In the inverse optimal control approach, a CLF candidate is used to construct an optimal control law directly without solving the associated HJB equation [36]. We focus on inverse optimality because it avoids solving the HJB partial differential equations and still allows us to obtain Kalman-type stability margins [56].

In contrast to the inverse optimal control via passivity, in which a storage function is used as a CLF candidate and the inverse optimal control law is selected as the output feedback, for the inverse optimal control via CLF, the control law is obtained as a result of solving the Bellman equation. Then, a CLF candidate for the obtained control law is proposed such that it stabilizes the system and a posteriori a cost functional is minimized.

For this control scheme, a quadratic CLF candidate is used to synthesize the inverse optimal control law. We establish the following assumptions and definitions which allow the inverse optimal control solution via the CLF approach.

Assumption 2 *The full state of system (2.1) is measurable.*

Along the lines of [121], we propose the discrete-time inverse optimal control law for nonlinear systems as follows.

DEFINITION 4.1: Inverse Optimal Control Law Let the control law

$$u_k^* = -\frac{1}{2}R^{-1}g^T(x_k)\frac{\partial V(x_{k+1})}{\partial x_{k+1}} \tag{4.1}$$

to be inverse optimal if

(i) it achieves (global) exponential stability of the equilibrium point $x_k = 0$ for system (2.1);

(ii) it minimizes a cost functional defined as (2.2), for which $l(x_k) := -\overline{V}$ with

$$\overline{V} := V(x_{k+1}) - V(x_k) + u_k^{*T}Ru_k^* \le 0. \tag{4.2}$$

As established in Definition 4.1, inverse optimal control is based on the knowledge of $V(x_k)$; thus, we propose a CLF based on $V(x_k)$ such that (i) and (ii) can be guaranteed. That is, instead of solving (2.10) for $V(x_k)$, we propose a control Lyapunov function $V(x_k)$ as

$$V(x_k) = \frac{1}{2} x_k^T P x_k, \qquad P = P^T > 0 \tag{4.3}$$

for control law (4.1) in order to ensure stability of system (2.1) equilibrium point $x_k = 0$, which will be achieved by defining an appropriate matrix P. Moreover, it will be established that control law (4.1) with (4.3), which is referred to as the inverse optimal control law, optimizes a cost functional of the form (2.2). Consequently, by considering $V(x_k)$ as in (4.3), control law (4.1) takes the following form:

$$
\begin{aligned}
u_k^* &= -\frac{1}{2} R^{-1} g^T(x_k) \frac{\partial V(x_{k+1})}{\partial x_{k+1}} \\
&= -\frac{1}{2} R^{-1} g^T(x_k)(P x_{k+1}) \\
&= -\frac{1}{2} R^{-1} g^T(x_k)(P f(x_k) + P g(x_k) u_k^*).
\end{aligned}
$$

Thus,

$$
\begin{aligned}
\left(I + \frac{1}{2} R^{-1} g^T(x_k) P g(x_k) \right) u_k^* &= \\
-\frac{1}{2} R^{-1} g^T(x_k) P f(x_k). &
\end{aligned}
\tag{4.4}
$$

Multiplying (4.4) by R, then

$$\left(R + \frac{1}{2} g^T(x_k) P g(x_k) \right) u_k^* = -\frac{1}{2} g^T(x_k) P f(x_k) \tag{4.5}$$

which results in the following state feedback control law:

$$\alpha(x_k) := u_k^*$$

$$= -\frac{1}{2}(R + P_2(x_k))^{-1} P_1(x_k) \tag{4.6}$$

where $P_1(x_k) = g^T(x_k) P f(x_k)$ and $P_2(x_k) = \frac{1}{2} g^T(x_k) P g(x_k)$. Note that $P_2(x_k)$ is a

positive definite and symmetric matrix, which ensures that the inverse matrix in (4.6)

exists.

Once we have proposed a CLF for solving the inverse optimal control in accordance

with Definition 4.1, the main contribution is presented as the following theorem.

Theorem 4.1

Consider the affine discrete-time nonlinear system (2.1). If there exists a matrix

$P = P^T > 0$ such that the following inequality holds

$$V_f(x_k) - \frac{1}{4} P_1^T(x_k)(R + P_2(x_k))^{-1}$$

$$\times P_1(x_k) \le -\zeta_Q \|x_k\|^2 . \tag{4.7}$$

where $V_f(x_k) = V(f(x_k)) - V(x_k)$, with $V(f(x_k)) = \frac{1}{2} f^T(x_k) P f(x_k)$ and $\zeta_Q > 0$;

$P_1(x_k)$ and $P_2(x_k)$ as defined in (4.6); then the equilibrium point $x_k = 0$ of system

(2.1) is globally exponentially stabilized by the control law (4.6), with CLF (4.3).

Moreover, with (4.3) as a CLF, this control law is inverse optimal in the sense that it minimizes the cost functional given by

$$V(x_k) = \sum_{k=0}^{\infty} \left(l(x_k) + u_k^T R u_k \right) \tag{4.8}$$

with

$$l(x_k) = -\overline{V}\big|_{u_k^* = \alpha(x_k)} \tag{4.9}$$

and optimal value function $V^*(x_0) = V(x_0)$.   ∎

PROOF

First, we analyze stability. Global stability for the equilibrium point $x_k = 0$ of system (2.1) with (4.6) as input is achieved if (4.2) is satisfied. Thus, $\overline{V}$ results in

$$
\begin{aligned}
\overline{V} &= V(x_{k+1}) - V(x_k) + \alpha^T(x_k) R \alpha(x_k) \\
&= \frac{f^T(x_k) P f(x_k) + 2 f^T(x_k) P g(x_k) \alpha(x_k)}{2} \\
&\quad + \frac{\alpha^T(x_k) g^T(x_k) P g(x_k) \alpha(x_k) - x_k^T P x_k}{2} + \alpha^T(x_k) R \alpha(x_k) \\
&= V_f(x_k) - \frac{1}{2} P_1^T(x_k) \left(R + P_2(x_k) \right)^{-1} P_1(x_k) \\
&\quad + \frac{1}{4} P_1^T(x_k) \left(R + P_2(x_k) \right)^{-1} P_1(x_k) \\
&= V_f(x_k) - \frac{1}{4} P_1^T(x_k) \left(R + P_2(x_k) \right)^{-1} P_1(x_k).
\end{aligned} \tag{4.10}
$$

Selecting P such that $\overline{V} \leq 0$, the stability of $x_k = 0$ is guaranteed. Furthermore, by means of P, we can achieve a desired negativity amount [37] for the closed-loop function $\overline{V}$ in (4.10). This negativity amount can be bounded using a positive definite matrix Q as follows:

$$
\begin{aligned}
\overline{V} &= V_f(x_k) - \frac{1}{4}P_1^T(x_k)(R+P_2(x_k))^{-1}P_1(x_k) \\
&\leq -x_k^T Q x_k \\
&\leq -\lambda_{min}(Q)\|x_k\|^2 \\
&= -\zeta_Q\|x_k\|^2, \qquad \zeta_Q = \lambda_{min}(Q) \qquad (4.11)
\end{aligned}
$$

where $\|\cdot\|$ stands for the Euclidean norm and $\zeta_Q > 0$ denotes the minimum eigenvalue

of matrix Q ($\lambda_{min}(Q)$). Thus, from (4.11) follows condition (4.7).

Considering (4.10)–(4.11)

$$\overline{V} = V(x_{k+1}) - V(x_k) + \alpha^T(x_k)R\alpha(x_k) \leq -\zeta_Q\|x_k\|^2 \Rightarrow \Delta V = V(x_{k+1}) - V(x_k) \leq$$
$$-\zeta_Q\|x_k\|^2.$$

Moreover, as $V(x_k)$ is a radially unbounded function, then the solution $x_k = 0$ of the

closed-loop system (2.1) with (4.6) as input is globally exponentially stable according

to Theorem 2.2.

When function $-l(x_k)$ is set to be the (4.11) right-hand side, that is,

$$
\begin{aligned}
l(x_k) &:= -\overline{V}\big|_{u_k^* = \alpha(x_k)} \qquad\qquad\qquad (4.12) \\
&= -V_f(x_k) + \frac{1}{4}P_1^T(x_k)(R+P_2(x_k))^{-1}P_1(x_k)
\end{aligned}
$$

then $V(x_k)$ as proposed in (4.3) is a solution of the DT HJB equation (2.10).

In order to obtain the optimal value for the cost functional (4.8), we substitute $l(x_k)$

given in (4.12) into (4.8); then

$$
\begin{aligned}
V(x_k) &= \sum_{k=0}^{\infty} \left(l(x_k) + u_k^T R u_k \right) \\
&= \sum_{k=0}^{\infty} \left(-\overline{V} + u_k^T R u_k \right) \\
&= -\sum_{k=0}^{\infty} \left[V_f(x_k) - \frac{1}{4} P_1^T(x_k) \left(R + P_2(x_k) \right)^{-1} P_1(x_k) \right] + \sum_{k=0}^{\infty} u_k^T R u_k.
\end{aligned}
\tag{4.13}
$$

Factorizing (4.13), and then adding the identity matrix

$$
I_m = \left(R + P_2(x_k) \right) \left(R + P_2(x_k) \right)^{-1}
$$

with $I_m \in \mathbb{R}^{m \times m}$, we obtain

$$
\begin{aligned}
V(x_k) = &-\sum_{k=0}^{\infty} \left[V_f(x_k) - \frac{1}{2} P_1^T(x_k) \left(R + P_2(x_k) \right)^{-1} P_1(x_k) \right. \\
&+ \frac{1}{4} P_1^T(x_k) \left(R + P_2(x_k) \right)^{-1} P_2(x_k) \left(R + P_2(x_k) \right)^{-1} P_1(x_k) \\
&+ \frac{1}{4} P_1^T(x_k) \left(R + P_2(x_k) \right)^{-1} R \\
&\left. \times \left(R + P_2(x_k) \right)^{-1} P_1(x_k) \right] + \sum_{k=0}^{\infty} u_k^T R u_k.
\end{aligned}
\tag{4.14}
$$

Being $\alpha(x_k) = -\frac{1}{2} \left(R + P_2(x_k) \right)^{-1} P_1(x_k)$, then (4.14) becomes

$$
\begin{aligned}
V(x_k) = &-\sum_{k=0}^{\infty} \left[V_f(x_k) + P_1^T(x_k) \alpha(x_k) + \alpha^T(x_k) P_2(x_k) \alpha(x_k) \right] + \sum_{k=0}^{\infty} \left[u_k^T R u_k \right. \\
&\left. - \alpha^T(x_k) R \alpha(x_k) \right] \\
= &-\sum_{k=0}^{\infty} \left[V(x_{k+1}) - V(x_k) \right] + \sum_{k=0}^{\infty} \left[u_k^T R u_k - \alpha^T(x_k) R \alpha(x_k) \right]
\end{aligned}
\tag{4.15}
$$

which can be written as

$$
\begin{aligned}
V(x_k) &= -\sum_{k=1}^{\infty}\left[V(x_{k+1})-V(x_k)\right]-V(x_1)+V(x_0) \\
&\quad +\sum_{k=0}^{\infty}\left[u_k^T R u_k-\alpha^T(x_k)R\alpha(x_k)\right] \\
&= -\sum_{k=2}^{\infty}\left[V(x_{k+1})-V(x_k)\right]-V(x_2)+V(x_1) \\
&\quad -V(x_1)+V(x_0)+\sum_{k=0}^{\infty}\left[u_k^T R u_k-\alpha^T(x_k)R\alpha(x_k)\right].
\end{aligned}
\tag{4.16}
$$

For notation convenience in (4.16), the upper limit ∞ will be treated as $N \to \infty$, and

thus

$$
\begin{aligned}
V(x_k) &= -V(x_N)+V(x_{N-1})-V(x_{N-1})+V(x_0) \\
&\quad +\lim_{N\to\infty}\sum_{k=0}^{N}\left[u_k^T R u_k-\alpha^T(x_k)R\alpha(x_k)\right] \\
&= -V(x_N)+V(x_0)+\lim_{N\to\infty}\sum_{k=0}^{N}\Big[u_k^T R u_k \\
&\quad -\alpha^T(x_k)R\alpha(x_k)\Big].
\end{aligned}
$$

Letting $N \to \infty$ and noting that $V(x_N) \to 0$ for all x_0, then

$$
V(x_k) = V(x_0)+\sum_{k=0}^{\infty}\left[u_k^T R u_k-\alpha^T(x_k)R\alpha(x_k)\right].
\tag{4.17}
$$

Thus, the minimum value of (4.17) is reached with $u_k = \alpha(x_k)$. Hence, the control

law (4.6) minimizes the cost functional (4.8). The optimal value function of (4.8) is

$V^*(x_k) = V(x_0)$ for all x_0.   ∎

REMARK 4.1 Additionally, with $l(x_k)$ as defined in (4.9), $V(x_k)$ solves the following

Hamilton–Jacobi–Bellman equation:

$$l(x_k) + V(x_{k+1}) - V(x_k) + \frac{1}{4} \frac{\partial V^T(x_{k+1})}{\partial x_{k+1}} g(x_k)$$
$$\times R^{-1} g^T(x_k) \frac{\partial V(x_{k+1})}{\partial x_{k+1}} = 0. \tag{4.18}$$

■

It can establish the main conceptual differences between optimal control and inverse optimal control as:

- For optimal control, the state cost function $l(x_k) \geq 0$ and the input weighing term $R > 0$ are given a priori. Then, they are used to determine $u(x_k)$ and $V(x_k)$ by means of the discrete-time HJB equation solution.

- For inverse optimal control, the control Lyapunov function $V(x_k)$ and the input weighting term R are given a priori. Then, these functions are used to compute $u(x_k)$ and $l(x_k)$ defined as $l(x_k) := -\bar{V}$.

Optimal control will be in general given as (4.1), and the minimum value of the cost functional (4.8) will be a function of the initial state x_0, that is, $V(x_0)$. If system (2.1) and the control law (4.1) are given, we shall say that the pair $\{V(x_k), l(x_k)\}$ is a solution to the *inverse optimal control* if the performance index (2.2) is minimized by (4.1), the minimum value being $V(x_0)$ [84].

4.1.1 EXAMPLE

The applicability of the developed method is illustrated by synthesis of a stabilizing control law for a discrete-time second order nonlinear system (unstable for $u_k = 0$)

of the form (2.1), with

$$f(x_k) = \begin{bmatrix} x_{1,k}x_{2,k} - 0.8x_{2,k} \\ x_{1,k}^2 + 1.8x_{2,k} \end{bmatrix} \qquad (4.19)$$

and

$$g(x_k) = \begin{bmatrix} 0 \\ -2 + cos(x_{2,k}) \end{bmatrix}. \qquad (4.20)$$

According to (4.6), the stabilizing optimal control law is formulated as

$$\alpha(x_k) = -\frac{1}{2}\left(R + \frac{1}{2}g^T(x_k)Pg(x_k)\right)^{-1}g^T(x_k)Pf(x_k)$$

where the positive definite matrix P is selected as

$$P = \begin{bmatrix} 10 & 0 \\ 0 & 10 \end{bmatrix}$$

and R is selected as the constant term $R = 1$.

The state penalty term $l(x_k)$ in (4.8) is calculated according to (4.9). The phase portrait for this unstable open-loop ($u_k = 0$) system with initial conditions $\chi_0 = [2 \quad -2]^T$ is displayed in Figure 4.1. Figure 4.2 presents the time evolution of x_k for this system with initial conditions $x_0 = [2 \quad -2]^T$ under the action of the proposed control law. This figure also includes the applied inverse optimal control law, which achieves stability; the respective phase portrait is displayed in Figure 4.3. Figure 4.4 displays the evaluation of the cost functional.

REMARK 4.2 For this example, according to Theorem 4.1, the optimal value function

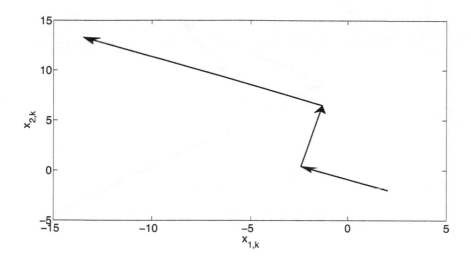

FIGURE 4.1   Unstable phase portrait.

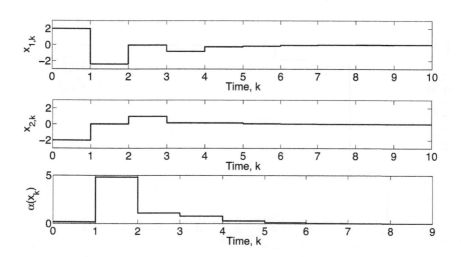

FIGURE 4.2   Stabilization of a nonlinear system.

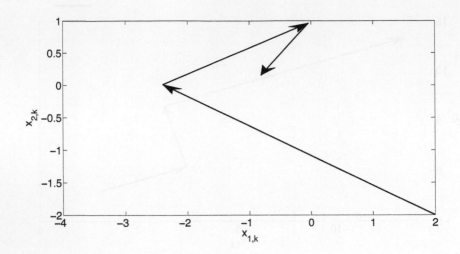

FIGURE 4.3   Phase portrait for the stabilized system.

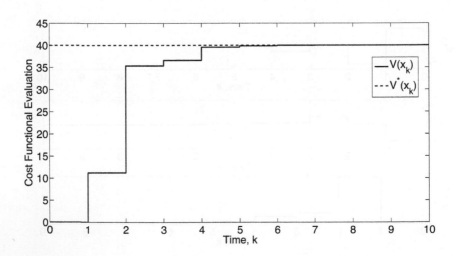

FIGURE 4.4   Cost functional evaluation.

is calculated as $V^*(x_k) = V(x_0) = \frac{1}{2} x_0^T P x_0 = 40$, which is reached as shown in Figure 4.4.

$\blacksquare$

4.1.2 INVERSE OPTIMAL CONTROL FOR LINEAR SYSTEMS

For the special case of linear systems, it can be shown that inverse optimal control is an alternative way to achieve stability and the minimization of a cost functional, avoiding the need to solve the discrete-time algebraic Riccati equation (DARE) [2]. That is, for the discrete-time linear system

$$x_{k+1} = A x_k + B u_k, \qquad x_0 = x(0) \tag{4.21}$$

the stabilizing inverse control law (4.6) becomes

$$
\begin{aligned}
u_k^* &= -\frac{1}{2} (R + P_2(x_k))^{-1} P_1(x_k) \\
&= -\frac{1}{2} (R + \frac{1}{2} B^T P B)^{-1} B^T P A x_k
\end{aligned}
\tag{4.22}
$$

where $P_1(x_k)$ and $P_2(x_k)$, by considering $f(x_k) = A x_k$ and $g(x_k) = B$, are defined as

$$P_1(x_k) = B^T P A x_k, \tag{4.23}$$

and

$$P_2(x_k) = \frac{1}{2} B^T P B. \tag{4.24}$$

Selecting $R = \frac{1}{2} \bar{R} > 0$, (4.22) results in

$$
\begin{aligned}
u_k^* &= -\frac{1}{2} (\frac{1}{2} \bar{R} + \frac{1}{2} B^T P B)^{-1} B^T P A x_k \\
&= -(\bar{R} + B^T P B)^{-1} B^T P A x_k.
\end{aligned}
\tag{4.25}
$$

For this linear case, function $\overline{V}$, as given in (4.10), becomes

$$
\begin{aligned}
\overline{V} &= V_f(x_k) - \frac{1}{4} P_1^T(x_k) \left(R + P_2(x_k)\right)^{-1} P_1(x_k) \\
&= \frac{x_k^T A^T P A x_k}{2} - \frac{x_k^T P x_k}{2} \\
&\quad - x_k^T A^T P B (\overline{R} + B^T P B)^{-1} B^T P A x_k \\
&\quad + \frac{u_k^{*T} B^T P B u_k^*}{2} + \frac{u_k^{*T} \overline{R}(x_k) u_k^*}{2}.
\end{aligned}
\tag{4.26}
$$

By means of P, we can achieve a desired negativity amount [37] for the function $\overline{V}$ in (4.26). This negativity amount can be bounded using a positive definite matrix $Q = \frac{1}{2}\overline{Q} > 0$ as follows:

$$
\overline{V} \le -\frac{x_k^T \overline{Q} x_k}{2}.
\tag{4.27}
$$

If we determine P such that (4.27) is satisfied, then the closed-loop system (4.21) with the inverse optimal control law (4.25) is globally exponentially stable. Note that condition (4.26)–(4.27) is the linear version of (4.7). Moreover, inequality (4.27) can be written as

$$
\begin{aligned}
x_k^T P x_k &= x_k^T \overline{Q} x_k + x_k^T A^T P A x_k \\
&\quad - 2 x_k^T A^T P B (\overline{R} + B^T P B)^{-1} B^T P A x_k \\
&\quad + x_k^T A^T P B (\overline{R} + B^T P B)^{-1} \\
&\quad \times (\overline{R} + B^T P B)(\overline{R} + B^T P B)^{-1} B^T P A x_k.
\end{aligned}
\tag{4.28}
$$

Finally, from (4.28) the discrete-time algebraic Riccati equation [2]

$$
P = \overline{Q} + A^T P A - A^T P B (\overline{R} + B^T P B)^{-1} B^T P A
\tag{4.29}
$$

is obtained, as in the case of the linear optimal regulator [8, 2].

The cost functional, which is minimized for the inverse optimal control law (4.25), results in

$$
\begin{aligned}
V(x_k) &= \sum_{k=0}^{\infty} \left(l(x_k) + \frac{1}{2} u_k^T \overline{R} u_k \right) \\
&= \frac{1}{2} \sum_{k=0}^{\infty} \left(u_k^T \overline{Q} u_k + u_k^T \overline{R} u_k \right)
\end{aligned}
\tag{4.30}
$$

where $l(x_k)$ is selected as $l(x_k) := -\overline{V} = \frac{1}{2} x_k^T \overline{Q} x_k$.

4.2 ROBUST INVERSE OPTIMAL CONTROL

Optimal controllers are known to be robust with respect to certain plant parameter variations, disturbances, and unmodeled dynamics as provided by stability margins, which implies that the Lyapunov difference $\Delta V < 0$ for optimal control schemes might still hold even for internal and/or external disturbances in the plant, and therefore stability will be maintained [81].

In this section, we establish a robust inverse optimal controller to achieve disturbance attenuation for a disturbed discrete-time nonlinear system. At the same time, this controller is optimal with respect to a cost functional, and we avoid solving the Hamilton–Jacobi–Isaacs (HJI) partial differential equation [36].

Let us consider the disturbed discrete-time nonlinear system

$$
x_{k+1} = f(x_k) + g(x_k) u_k + d_k, \qquad x_0 = x(0)
\tag{4.31}
$$

where $d_k \in \mathbb{R}^n$ is a disturbance which is bounded by

$$
\| d_k \| \leq \ell_k' + \alpha_4(\|x_k\|)
\tag{4.32}
$$

with $\ell'_k \le \ell$; ℓ is a positive constant and $\alpha_4(\|x_k\|)$ is a $\mathcal{K}_\infty$– function, and suppose that

$\alpha_4(\|x_k\|)$ in (4.32) is of the same order as the $\mathcal{K}_\infty$– function $\alpha_3(\|x_k\|)$, i.e.,

$$\alpha_4(\|x_k\|) = \delta\, \alpha_3(\|x_k\|), \qquad \delta > 0. \tag{4.33}$$

In the next definition, we establish the discrete-time *robust inverse optimal* control.

DEFINITION 4.2: Robust Inverse Optimal Control Law The control law

$$u^*_k = \alpha(x_k) = -\frac{1}{2}R^{-1}g^T(x_k)\frac{\partial V(x_{k+1})}{\partial x_{k+1}} \tag{4.34}$$

is robust inverse optimal if

(i) it achieves (global) ISS for system (4.31);

(ii) $V(x_k)$ is (radially unbounded) positive definite such that the inequality

$$\overline{V}_d(x_k, d_k): \quad = \quad V(x_{k+1}) - V(x_k) + u_k^T R u_k$$

$$\le \quad -\sigma(x_k) + \ell_d \|d_k\| \tag{4.35}$$

is satisfied, where $\sigma(x_k)$ is a positive definite function and ℓ_d is a positive constant.

The value of function $\sigma(x_k)$ represents a desired amount of negativity [37] of the

closed-loop Lyapunov difference $\overline{V}_d(x_k, d_k)$.

For the robust inverse optimal control solution, let us consider the continuous

state feedback control law (4.34), with (4.3) as a CLF candidate, where $P \in \mathbb{R}^{n \times n}$ is

assumed to be a positive definite and symmetric matrix. Taking one step ahead for

(4.3), then control law (4.34) results in (4.6).

Hence, a robust inverse optimal controller is stated as follows.

Theorem 4.2

Consider a disturbed affine discrete-time nonlinear system (4.31). If there exists a matrix $P = P^T > 0$ such that the following inequality holds

$$V_f(x_k) - \frac{1}{4} P_1^T(x_k)(R + P_2(x_k))^{-1} P_1(x_k) \leq -\zeta \alpha_3(\|x_k\|), \quad \forall \|x_k\| \geq \rho(\|d_k\|) \quad (4.36)$$

with δ in (4.33) satisfying

$$\delta < \frac{\eta}{\ell_d} \quad (4.37)$$

where function $V_f(x_k) = V(f(x_k)) - V(x_k)$, and with $P_1(x_k) = g^T(x_k) P f(x_k)$ and $P_2(x_k) = \frac{1}{2} g^T(x_k) P g(x_k)$; ζ, $\ell_d > 0$, $\eta = (1 - \theta)\zeta > 0$, $0 < \theta < 1$, and with ρ a $\mathcal{K}_\infty-$ function, then the solution of the closed-loop system (4.31) and (4.6) is ISS with the ultimate bound γ (i.e., $\|x_k\| \leq \gamma$, $\forall k \geq k_0 + T$) and (4.3) as an ISS–CLF in (2.23)–(2.24). The ultimate bound γ in (2.17) becomes $\gamma = \alpha_3^{-1}\left(\frac{\ell_d \ell}{\theta_1 \zeta}\right)\sqrt{\frac{\lambda_{max}(P)}{\lambda_{min}(P)}}$.

Moreover, with (4.3) as an ISS–CLF, control law (4.6) is inverse optimal in the sense that it minimizes the cost functional given as

$$\mathcal{I} = \sup_{d \in \mathcal{D}} \left\{ \lim_{\tau \to \infty} \left[V(x_\tau) + \sum_{k=0}^{\tau} \left(l_d(x_k) + u_k^T R u_k + \ell_d \|d_k\| \right) \right] \right\} \quad (4.38)$$

where $\mathcal{D}$ is the set of locally bounded functions, and

$$l_d(x_k) := -V_d(x_k, d_k)$$

with $V_d(x_k, d_k)$ a negative definite function.   ∎

PROOF

First, we analyze stability for system (4.31) with nonvanishing disturbance d_k. It

is worth noting that asymptotic stability of $x = 0$ is not reached anymore [49]; the

ISS property for solution of system (4.31) can only be ensured if stabilizability is

assumed. Stability analysis for a disturbed system can be treated by two terms; we

propose a Lyapunov difference for the nominal system (i.e., $x_{k+1} = f(x_k) + g(x_k)u_k$),

denoted by ΔV, and additionally, a difference for disturbed system (4.31) denoted by

ΔV. The Lyapunov difference for the disturbed system is defined as

$$\Delta V_d(x_k, d_k) = V(x_{k+1}) - V(x_k). \tag{4.39}$$

Let us first define the function $V_{nom}(x_{k+1})$ as the $k+1$–step using the Lyapunov

function $V(x_k)$ for the nominal system (2.1). Then, adding and subtracting $V_{nom}(x_{k+1})$

in (4.39)

$$\Delta V_d(x_k, d_k) = \underbrace{V(x_{k+1}) - V_{nom}(x_{k+1})}_{\Delta V :=} + \underbrace{V_{nom}(x_{k+1}) - V(x_k)}_{\Delta V :=}. \tag{4.40}$$

From (4.35) with $\sigma(x_k) = \zeta \alpha_3(\|x_k\|)$, $\zeta > 0$, and the control law (4.6), we obtain

$$\begin{aligned} \Delta V &= V_f(x_k) - \frac{1}{4}P_1^T(x_k)(R + P_2(x_k))^{-1}P_1(x_k) \\ &\leq -\zeta \alpha_3(\|x_k\|) \end{aligned}$$

in (4.40), which is ensured by means of $P = P^T > 0$. On the other hand, since $V(x_k)$ is

a C^1 (indeed it is C^2 differentiable) function in x_k for all k, then ΔV satisfies condition

(2.19) as

$$|\Lambda V| \leq \ell_d \| f(x_k) + g(x_k) u_k(x_k) + d_k - f(x_k) - g(x_k) u_k(x_k) \|$$

$$= \ell_d \| d_k \|$$

$$\leq \ell_d \ell + \ell_d \alpha_4(\|x_k\|)$$

where ℓ and ℓ_d are positive constants. Hence, the Lyapunov difference $\Delta V_d(x_k, d_k)$ for the disturbed system is determined as

$$\Delta V_d(x_k, d_k) = \Lambda V + \Delta V$$

$$\leq |\Lambda V| + \Delta V$$

$$\leq \ell_d \alpha_4(\|x_k\|) + \ell_d \ell + V_f(x_k) - \frac{1}{4} P_1^T(x_k)(R + P_2(x_k))^{-1} P_1(x_k)$$

$$\leq -\zeta \alpha_3(\|x_k\|) + \ell_d \alpha_4(\|x_k\|) + \ell_d \ell \qquad (4.41)$$

$$= -\zeta \alpha_3(\|x_k\|) + \theta \zeta \alpha_3(\|x_k\|) - \theta \zeta \alpha_3(\|x_k\|) + \ell_d \alpha_4(\|x_k\|) + \ell_d \ell$$

$$= -(1 - \theta) \zeta \alpha_3(\|x_k\|) - \theta \zeta \alpha_3(\|x_k\|) + \ell_d \alpha_4(\|x_k\|) + \ell_d \ell$$

$$= -(1 - \theta) \zeta \alpha_3(\|x_k\|) + \ell_d \alpha_4(\|x_k\|), \qquad \forall \|x_k\| \geq \alpha_3^{-1}\left(\frac{\ell_d \ell}{\theta \zeta}\right)$$

where $0 < \theta < 1$. In particular, using condition (4.33) in the previous expression, we obtain

$$\Delta V_d(x_k, d_k) \leq -(1 - \theta) \zeta \alpha_3(\|x_k\|) + \ell_d \alpha_4(\|x_k\|)$$

$$= -\eta \alpha_3(\|x_k\|) + \ell_d \delta \alpha_3(\|x_k\|) \qquad (4.42)$$

$$= -(\eta - \ell_d \delta) \alpha_3(\|x_k\|), \qquad \eta = (1 - \theta) \zeta > 0$$

which is negative definite if condition $\delta < \frac{\eta}{\ell_d}$ (4.37) is satisfied. Therefore, if condition (4.37) holds and considering $V(x_k)$ (4.3) as a radially unbounded ISS–CLF, then by

Proposition 2.2, the closed-loop system (4.31) and (4.6) is ISS, which implies BIBS

stability and $\mathcal{K}$– asymptotic gain according to Theorem 2.3.

By Definition 2.11 and Remark 2.1, the solution of the closed-loop system

(4.31) and (4.6) is ultimately bounded with $\gamma = \alpha_1^{-1} \circ \alpha_2 \circ \rho$, which results in

$\gamma = \alpha_3^{-1}\left(\frac{\ell_d \ell}{\theta \zeta}\right)\sqrt{\frac{\lambda_{max}(P)}{\lambda_{min}(P)}}$. Hence, according to Definition 2.4, the solution is ulti-

mately bounded with ultimate bound $b = \gamma$.

In order to establish inverse optimality, considering that the control (4.6) achieves

ISS for the system (4.31), and substituting $l_d(x_k)$ in (4.38), it follows that

$$
\begin{aligned}
\mathscr{J} &= \sup_{d\in\mathscr{D}}\left\{\lim_{\tau\to\infty}\left[V(x_\tau)+\sum_{k=0}^{\tau}\left(l_d(x_k)+u_k^T R u_k+\ell_d\,\|d_k\|\right)\right]\right\}\\
&= \sup_{d\in\mathscr{D}}\left\{\lim_{\tau\to\infty}\left[V(x_\tau)+\sum_{k=0}^{\tau}\left(-\Lambda V-\Delta V+u_k^T R u_k+\ell_d\,\|d_k\|\right)\right]\right\}\\
&= \sup_{d\in\mathscr{D}}\left\{\lim_{\tau\to\infty}\left[V(x_\tau)-\sum_{k=0}^{\tau}\left(V_f(x_k)-\frac{1}{4}P_1^T(x_k)(R+P_2(x_k))^{-1}P_1(x_k)\right.\right.\right.\\
&\qquad\left.\left.\left.+\ell_d\ell+\ell_d\delta\alpha_3(\|x_k\|)\right)+\sum_{k=0}^{\tau}u_k^T R u_k+\sum_{k=0}^{\tau}\ell_d\,\|d_k\|\right]\right\}\\
&= \lim_{\tau\to\infty}\left[V(x_\tau)-\sum_{k=0}^{\tau}\left(V_f(x_k)-\frac{1}{4}P_1^T(x_k)(R+P_2(x_k))^{-1}P_1(x_k)\right)\right.\\
&\qquad\left.+\sum_{k=0}^{\tau}u_k^T R u_k+\sup_{d\in\mathscr{D}}\left\{\sum_{k=0}^{\tau}\left(\ell_d\,\|d_k\|-\ell_d\ell-\ell_d\delta\alpha_3(\|x_k\|)\right)\right\}\right]. \quad (4.43)
\end{aligned}
$$

Adding the term

$$
\frac{1}{4}P_1^T(x_k)(R+P_2(x_k))^{-1}R(R+P_2(x_k))^{-1}P_1(x_k)
$$

at the first addition term of (4.43) and subtracting at the second addition term of (4.43) yields

$$
\begin{aligned}
\mathscr{J} &= \lim_{\tau \to \infty} \Bigg[V(x_\tau) - \sum_{k=0}^{\tau} (V(x_{k+1}) - V(x_k)) + \sum_{k=0}^{\tau} \Bigg(u_k^T R u_k - \frac{1}{4} P_1^T (x_k) (R \\
&\quad + P_2(x_k))^{-1} R (R + P_2(x_k))^{-1} P_1(x_k) \Bigg) \\
&\quad + \sup_{d \in \mathscr{D}} \Bigg\{ \sum_{k=0}^{\tau} (\ell_d \|d_k\| - \ell_d \ell - \ell_d \delta \alpha_3 (\|x_k\|)) \Bigg\} \Bigg] \\
&= \lim_{\tau \to \infty} \Bigg[V(x_\tau) - \sum_{k=0}^{\tau} (V(x_{k+1}) - V(x_k)) + \sum_{k=0}^{\tau} [u_k^T R u_k - \alpha^T (x_k) R \alpha(x_k)] \\
&\quad + \sup_{d \in \mathscr{D}} \Bigg\{ \sum_{k=0}^{\tau} (\ell_d \|d_k\| - \ell_d \ell - \ell_d \delta \alpha_3 (\|x_k\|)) \Bigg\} \Bigg] \\
&= \lim_{\tau \to \infty} \Bigg[V(x_\tau) - V(x_\tau) + V(x_0) + \sum_{k=0}^{\tau} [u_k^T R u_k - \alpha^T (x_k) R \alpha(x_k)] \\
&\quad + \sum_{k=0}^{\tau} \Bigg(\sup_{d \in \mathscr{D}} \{\ell_d \|d_k\|\} - \ell_d \ell - \ell_d \delta \alpha_3 (\|x_k\|) \Bigg) \Bigg]. \quad (4.44)
\end{aligned}
$$

If $\sup_{d \in \mathscr{D}} \{\ell_d \|d_k\|\}$ is taken as the worst case by considering the equality for (4.32), we obtain

$$
\begin{aligned}
\sup_{d \in \mathscr{D}} \{\ell_d \|d_k\|\} &= \ell_d \sup_{d \in \mathscr{D}} \{\|d_k\|\} \\
&= \ell_d \ell + \ell_d \delta \alpha_3 (\|x_k\|). \quad (4.45)
\end{aligned}
$$

Therefore

$$
\sum_{k=0}^{\tau} \Bigg(\sup_{d \in \mathscr{D}} \{\ell_d \|d_k\|\} - \ell_d \ell - \ell_d \delta \alpha_3 (\|x_k\|) \Bigg) = 0. \quad (4.46)
$$

Thus, the minimum value of (4.44) is reached with $u_k = \alpha(x_k)$. Hence, the control law (4.6) minimizes the cost functional (4.38). The optimal value function of (4.38) is $\mathscr{J}^*(x_0) = V(x_0)$. ∎

REMARK 4.3 It is worth noting that, in the inverse optimality analysis proof, equality for (4.33) is considered in order to optimize with respect to the worst case for the disturbance.   ∎

As a special case, the manipulation of class $\mathcal{K}_\infty-$ functions in Definition 2.10 is simplified when the class $\mathcal{K}_\infty-$ functions take the special form $\alpha_i(r) = \kappa_i r^c$, $\kappa_i > 0$, $c > 1$, and $i = 1, 2, 3$. In this case, exponential stability is achieved [49]. Let us assume that the disturbance term d_k in (4.31) satisfies the bound

$$\|d_k\| \leq \ell + \delta \|x_k\|^2 \tag{4.47}$$

where ℓ and δ are positive constants, then the following result is stated.

COROLLARY 4.1

Consider the disturbed affine discrete-time nonlinear system (4.31) with (4.47). If there exists a matrix $P = P^T > 0$ such that the following inequality holds

$$V_f(x_k) - \frac{1}{4}P_1^T(x_k)(R + P_2(x_k))^{-1} P_1(x_k) \leq -\zeta_Q \|x_k\|^2 \quad \forall \|x_k\| \geq \rho(\|d_k\|) \tag{4.48}$$

where $\zeta_Q > 0$ denotes the minimum eigenvalue of matrix Q as established in (4.11), and δ in (4.47) satisfies

$$\delta < \frac{\zeta_Q}{\ell_d} \tag{4.49}$$

then the solution of closed-loop system (4.31), (4.6) is ISS, with (4.3) as an ISS–CLF. The ultimate bound γ in (2.17) becomes $\gamma = \sqrt{\frac{\ell_d \ell}{\theta \eta}} \sqrt{\frac{\lambda_{max}(P)}{\lambda_{min}(P)}}$ with $0 < \theta < 1$ and $\eta > 0$. This bound is reached exponentially.

Moreover, with (4.3) as an ISS–CLF, this control law is inverse optimal in the sense that it minimizes the cost functional given as

$$\mathscr{I} = \sup_{d \in \mathscr{D}} \left\{ \lim_{\tau \to \infty} \left[V(x_\tau) + \sum_{k=0}^{\tau} \left(l_d(x_k) + u_k^T R u_k + \ell_d \|d_k\| \right) \right] \right\} \qquad (4.50)$$

where $\mathscr{D}$ is the set of locally bounded functions, and

$$l_d(x_k) = -\Lambda V - \Delta V.$$

PROOF

Stability is analyzed similarly to the proof of Theorem 4.2, where the Lyapunov difference is treated by means of two terms as in (4.40). For the first one, the disturbance term is considered, and for the second one, the Lyapunov difference in order to achieve exponential stability for an undisturbed system is analyzed. For the latter, Lyapunov difference ΔV is considered from (4.11); hence the Lyapunov difference ΔV becomes $\Delta V \le -\zeta_Q \|x_k\|^2$ with a positive constant ζ_Q, and since $V(x_k)$ is a C^1 function in x_k for all k, then ΛV satisfies the bound condition (2.19) as

$$
\begin{aligned}
|\Lambda V| &\le \ell_d \|f(x_k) + g(x_k) u_k + d_k - f(x_k) + g(x_k) u_k\| \\
&= \ell_d \|d_k\| \\
&\le \ell_d \ell + \ell_d \delta \|x_k\|^2 \qquad (4.51)
\end{aligned}
$$

where ℓ_d and δ are positive constants, and the bound disturbance (4.47) is regarded. Hence, from (4.11) and (4.51) the Lyapunov difference $\Delta V_d(x_k, d_k)$ for disturbed system (4.31) is established as

$$
\begin{aligned}
\Delta V_d(x_k, d_k) &= \Lambda V + \Delta V \\
&\leq |\Lambda V| + \Delta V \\
&\leq \ell_d \, \delta \, \|x_k\|^2 + \ell_d \, \ell + V_f(x_k) - \frac{1}{4} P_1^T(x_k)\,(R + P_2(x_k))^{-1}\,P_1(x_k) \\
&\leq -x_k^T Q x_k + \ell_d \, \delta \, \|x_k\|^2 + \ell_d \, \ell \\
&\leq -\zeta_Q \|x_k\|^2 + \ell_d \, \delta \, \|x_k\|^2 + \ell_d \, \ell \\
&= -(\zeta_Q - \ell_d \delta)\,\|x_k\|^2 + \ell_d \, \ell \\
&= -\eta \, \|x_k\|^2 + \ell_d \, \ell, \qquad\qquad\qquad\qquad \eta = \zeta_Q - \ell_d \, \delta > 0 \\
&= -\eta \, \|x_k\|^2 + \theta \eta \, \|x_k\|^2 - \theta \eta \, \|x_k\|^2 + \ell_d \, \ell, \qquad 0 < \theta < 1 \\
&= -(1 - \theta)\eta \, \|x_k\|^2, \qquad\qquad\qquad \|x_k\| > \sqrt{\frac{\ell_d \, \ell}{\theta \eta}}. \qquad (4.52)
\end{aligned}
$$

At this point, it must be ensured that η in (4.52) is positive, and thus $\delta < \frac{\zeta_Q}{\ell_d}$, i.e., the

condition (4.49).

To this end, as $V(x_k)$ is a radially unbounded function ISS–CLF, then, by Proposi-

tion 2.2, the solution of the closed-loop system (4.6), (4.31) is ISS with exponential

convergence to the ultimate bound γ, which results in $\gamma = \sqrt{\frac{\ell_d \, \ell}{\theta \eta}} \, \sqrt{\frac{\lambda_{max}(P)}{\lambda_{min}(P)}}$.

In order to establish inverse optimality, considering that (4.6) achieves ISS for

(4.31), and substituting $l_d(x_k)$ in (4.50), it follows that

$$
\begin{aligned}
\mathscr{I} &= \sup_{d\in\mathscr{D}}\left\{\lim_{\tau\to\infty}\left[V(x_\tau)+\sum_{k=0}^{\tau}\left(l_d(x_k)+u_k^T R u_k+\ell_d\,\|d_k\|\right)\right]\right\}\\
&= \sup_{d\subset\mathscr{D}}\left\{\lim_{\tau\to\infty}\left[V(x_\tau)+\sum_{k=0}^{\tau}\left(-\Lambda V-\Delta V+u_k^T R u_k+\ell_d\,\|d_k\|\right)\right]\right\}\\
&= \sup_{d\in\mathscr{D}}\left\{\lim_{\tau\to\infty}\left[V(x_\tau)-\sum_{k=0}^{\tau}\left(V_f(x_k)-\frac{1}{4}P_1^T(x_k)\,(R+P_2(x_k))^{-1}P_1(x_k)\right.\right.\right.\\
&\qquad\left.\left.\left.+\ell_d\,\ell+\ell_d\delta\,\|x_k\|^2\right)+\sum_{k=0}^{\tau}u_k^T R u_k+\sum_{k=0}^{\tau}\ell_d\,\|d_k\|\right]\right\}\\
&= \lim_{\tau\to\infty}\left[V(x_\tau)-\sum_{k=0}^{\tau}\left(V_f(x_k)-\frac{1}{4}P_1^T(x_k)\,(R+P_2(x_k))^{-1}P_1(x_k)\right)\right.\\
&\qquad\left.+\sum_{k=0}^{\tau}u_k^T R u_k+\sup_{d\in\mathscr{D}}\left\{\sum_{k=0}^{\tau}\left(\ell_d\,\|d_k\|-\ell_d\,\ell-\ell_d\delta\,\|x_k\|^2\right)\right\}\right].
\end{aligned}
$$

Adding the term $\frac{1}{4}P_1^T(x_k)\,(R+P_2(x_k))^{-1}R\,(R+P_2(x_k))^{-1}P_1(x_k)$ to the first addition

term and subtracting in the second addition term yields

$$
\begin{aligned}
\mathscr{I} &= \lim_{\tau\to\infty}\left[V(x_\tau)-\sum_{k=0}^{\tau}(V(x_{k+1})-V(x_k))+\sum_{k=0}^{\tau}\left(u_k^T R u_k\right.\right.\\
&\qquad\left.-\frac{1}{4}P_1^T(x_k)\,(R+P_2(x_k))^{-1}R\,(R+P_2(x_k))^{-1}P_1(x_k)\right)\\
&\qquad\left.+\sup_{d\in\mathscr{D}}\left\{\sum_{k=0}^{\tau}\left(\ell_d\,\|d_k\|-\ell_d\,\ell-\ell_d\delta\,\|x_k\|^2\right)\right\}\right]\\
&= \lim_{\tau\to\infty}\left[V(x_\tau)-\sum_{k=0}^{\tau}(V(x_{k+1})-V(x_k))+\sum_{k=0}^{\tau}\left[u_k^T R u_k\right.\right.\\
&\qquad\left.-\alpha^T(x_k)R\alpha(x_k)\right]+\sup_{d\in\mathscr{D}}\left\{\sum_{k=0}^{\tau}\left(\ell_d\,\|d_k\|-\ell_d\,\ell-\ell_d\delta\,\|x_k\|^2\right)\right\}\right]\\
&= \lim_{\tau\to\infty}\left[V(x_\tau)-V(x_\tau)+V(x_0)+\sum_{k=0}^{\tau}\left[u_k^T R u_k-\alpha^T(x_k)R\alpha(x_k)\right]\right.\\
&\qquad\left.+\sum_{k=0}^{\tau}\left(\sup_{d\in\mathscr{D}}\{\ell_d\,\|d_k\|\}-\ell_d\,\ell-\ell_d\delta\,\|x_k\|^2\right)\right].
\end{aligned}
\tag{4.53}
$$

If $\sup_{d\in\mathscr{D}}\{\ell_d\,\|d_k\|\}$ is taken as the worst case by considering the equality for (4.47), we obtain

$$
\begin{aligned}
\sup_{d \in \mathscr{D}} \{ \ell_d \, \|d_k\| \} &= \ell_d \sup_{d \in \mathscr{D}} \{ \|d_k\| \} \\
&= \ell_d \ell + \ell_d \delta \, \|x_k\|^2 .
\end{aligned}
\tag{4.54}
$$

Therefore

$$
\sum_{k=0}^{\tau} \left(\sup_{d \in \mathscr{D}} \{ \ell_d \, \|d_k\| \} - \ell_d \ell - \ell_d \delta \, \|x_k\|^2 \right) = 0.
\tag{4.55}
$$

Thus, the minimum value of (4.53) is reached with $u_k = \alpha(x_k)$, and the control law

(4.6) minimizes the cost functional (4.50). The optimal value function of (4.50) is

$\mathscr{J}^*(x_0) = V(x_0)$. ∎

REMARK 4.4 Terminal penalty $V(x_\tau)$ in (4.38) and (4.50) is a necessary function

to avoid imposing the assumption $x_\tau \to 0$ as $\tau \to \infty$. Hence, inverse optimality is

guaranteed only outside the ball, which is bounded by function γ as defined in (2.17).

∎

4.3   TRAJECTORY TRACKING INVERSE OPTIMAL CONTROL

Consider the affine discrete-time nonlinear system (2.1). The following cost functional

is associated with trajectory tracking for system (2.1):

$$
\mathscr{J}(z_k) = \sum_{n=k}^{\infty} \left(l(z_n) + u_n^T R u_n \right)
\tag{4.56}
$$

where $z_k = x_k - x_{\delta,k}$ with $x_{\delta,k}$ as the desired trajectory for x_k; $z_k \in \mathbb{R}^n$; $\mathscr{J}(z_k) : \mathbb{R}^n \to$

$\mathbb{R}^+$; $l(z_k) : \mathbb{R}^n \to \mathbb{R}^+$ is a positive semidefinite function and $R : \mathbb{R}^n \to \mathbb{R}^{m \times m}$ is a real symmetric positive definite weighting matrix. The cost functional (4.56) is a performance measure [51]. The entries of R can be fixed or functions of the system state in order to vary the weighting on control efforts according to the state value [51]. Considering the state feedback control design problem, we assume that the full state x_k is available.

Using the optimal value function $\mathscr{J}^*(x_k)$ for (4.56) as Lyapunov function $V(x_k)$, equation (4.56) can be rewritten as

$$
\begin{aligned}
V(z_k) &= l(z_k) + u_k^T R u_k + \sum_{n=k+1}^{\infty} l(z_n) + u_n^T R u_n \\
&= l(z_k) + u_k^T R u_k + V(z_{k+1})
\end{aligned}
\tag{4.57}
$$

where we require the boundary condition $V(0) = 0$ so that $V(z_k)$ becomes a Lyapunov function.

Similar to the Section 2.1 procedure, we establish the conditions that the optimal control law must satisfy. We define the discrete-time Hamiltonian $\mathscr{H}(z_k, u_k)$ as

$$
\mathscr{H}(z_k, u_k) = l(z_k) + u_k^T R u_k + V(z_{k+1}) - V(z_k).
\tag{4.58}
$$

A necessary condition that the optimal control law should satisfy is $\frac{\partial \mathscr{H}(z_k, u_k)}{\partial u_k} = 0$, then

$$
\begin{aligned}
0 &= 2 R u_k + \frac{\partial V(z_{k+1})}{\partial u_k} \\
&= 2 R u_k + \frac{\partial z_{k+1}}{\partial u_k} \frac{\partial V(z_{k+1})}{\partial z_{k+1}} \\
&= 2 R u_k + g^T(x_k) \frac{\partial V(z_{k+1})}{\partial z_{k+1}}.
\end{aligned}
\tag{4.59}
$$

Therefore, the optimal control law to achieve trajectory tracking is formulated as

$$u_k^* = -\frac{1}{2}R^{-1}g^T(x_k)\frac{\partial V(z_{k+1})}{\partial z_{k+1}} \tag{4.60}$$

with the boundary condition $V(0) = 0$. For determining the trajectory tracking inverse

optimal control, it is necessary to solve the following HJB equation:

$$l(z_k) + V(z_{k+1}) - V(z_k) + \frac{1}{4}\frac{\partial V^T(z_{k+1})}{\partial z_{k+1}}g(x_k)R^{-1}g^T(x_k)\frac{\partial V(z_{k+1})}{\partial z_{k+1}} = 0 \tag{4.61}$$

which is a challenging task. To overcome this problem, we propose using inverse

optimal control. The main characteristic of inverse optimal control is that a stabilizing

feedback control law is designed first, and then it is established that this law optimizes

the cost functional (4.56).

DEFINITION 4.3: Tracking Inverse Optimal Control Law Consider the tracking

error as $z_k = x_k - x_{\delta,k}$, $x_{\delta,k}$ being the desired trajectory for x_k. Let us define the control

law

$$u_k^* = -\frac{1}{2}R^{-1}g^T(x_k)\frac{\partial V(z_{k+1})}{\partial z_{k+1}} \tag{4.62}$$

which will be inverse optimal stabilizing along the desired trajectory $x_{\delta,k}$ if

(i) it achieves (global) asymptotic stability of $x_k = 0$ for system (2.1) along reference

   $x_{\delta,k}$;

(ii) $V(z_k)$ is a (radially unbounded) positive definite function such that inequality

$$\overline{V} := V(z_{k+1}) - V(z_k) + u_k^{*T}Ru_k^* \leq 0 \tag{4.63}$$

is satisfied.

When we select $l(z_k) := -\bar{V}$, then $V(z_k)$ is a solution for (4.61), and cost functional (4.56) is minimized.

As established in Definition 4.3, the inverse optimal control law for trajectory tracking is based on the knowledge of $V(z_k)$. Then, we propose a CLF, $V(z_k)$, such that (i) and (ii) are guaranteed. Hence, instead of solving (4.61) for $V(z_k)$, a quadratic CLF candidate $V(z_k)$ for (4.62) is proposed with the form

$$V(z_k) = \frac{1}{2} z_k^T P z_k, \qquad P = P^T > 0 \qquad (4.64)$$

in order to ensure stability of the tracking error z_k, where

$$
\begin{aligned}
z_k &= x_k - x_{\delta,k} \\
&= \begin{bmatrix} (x_{1,k} - x_{1\delta,k}) \\ (x_{2,k} - x_{2\delta,k}) \\ \vdots \\ (x_{n,k} - x_{n\delta,k}) \end{bmatrix}.
\end{aligned}
\qquad (4.65)
$$

Moreover, it will be established that the control law (4.62) with (4.64), which is referred to as the *inverse optimal* control law, optimizes a cost functional of the form (4.56).

Consequently, by considering $V(x_k)$ as in (4.64), control law (4.62) takes the following form:

$$
\begin{aligned}
u_k^* &= -\frac{1}{4} R g^T(x_k) \frac{\partial z_{k+1}^T P z_{k+1}}{\partial z_{k+1}} \\
&= -\frac{1}{2} R g^T(x_k) P z_{k+1} \\
&= -\frac{1}{2} \left(R + \frac{1}{2} g^T(x_k) P g(x_k) \right)^{-1} g^T(x_k) P (f(x_k) - x_{\delta,k+1}).
\end{aligned}
\qquad (4.66)
$$

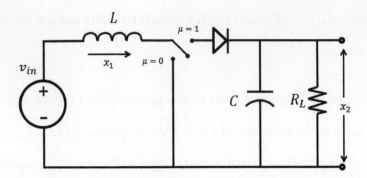

FIGURE 4.5   Boost converter circuit.

It is worth pointing out that P_k and R are positive definite and symmetric matrices; thus, the existence of the inverse in (4.66) is ensured.

4.3.1   APPLICATION TO THE BOOST CONVERTER

In this section, the inverse optimal control approach to achieve trajectory tracking, as described in the previous section, is used to control the capacitor voltage for a boost converter. Figure 4.5 shows the basic electrical circuit of the converter and the DC-to-DC converter is displayed in Figure 4.6.

4.3.1.1   Boost Converter Model

The commutated model for the boost converter can be presented as

$$\dot{x}_1 = -\frac{x_2}{L}\mu + \frac{v_{in}}{L}$$
$$\dot{x}_2 = -\frac{x_2}{R_L C} + \frac{x_1}{C}\mu \tag{4.67}$$

where x_1 is the current across the inductor, x_2 is the voltage in the capacitor, and $\mu = \{0,1\}$ define the switch position; the parameters $R_L, L, C,$ and v_{in} for the circuit

FIGURE 4.6 **(SEE COLOR INSERT)**   DC-to-DC boost converter.

are the resistance, inductance, capacitance, and source voltage, respectively, which

are assumed known.

It is well-known [47] that when the switching frequency is high, system (4.67) can

be represented by an *average model* given as

$$\dot{x}_1 = -\frac{x_2}{L}u + \frac{v_{in}}{L}$$
$$\dot{x}_2 = -\frac{x_2}{R_L C} + \frac{x_1}{C}u \tag{4.68}$$

where the control input $u = [0,1]$ is the duty cycle. The value of the duty cycle

determines the time in the pulse-width modulation (PWM) scheme, for which the

switch is fixed at the position represented by $\mu = 1$ (see Figure 4.5) [96]. Variables

x_1 and x_2 for (4.68) represent average values for current and voltage, respectively.

After discretizing by Euler approximation, the discrete-time model for the boost converter is rewritten as

$$x_{1,k+1} = x_{1,k} + T\left(\frac{v_{in}}{L} - \frac{x_{2,k}}{L}u_k\right)$$

$$x_{2,k+1} = x_{2,k} + T\left(-\frac{x_{2,k}}{R_L C} + \frac{x_{1,k}}{C}u_k\right) \tag{4.69}$$

where T is the sampling time. System (4.69) has the general affine form as (2.1) with

$$x_k = [x_{1,k}, x_{2,k}]^T, \quad f(x_k) = \begin{bmatrix} x_{1,k} + T\,v_{in}/L \\ \\ x_{2,k} - T\,x_{2,k}/R_L C \end{bmatrix} \quad \text{and} \quad g(x_k) = \begin{bmatrix} -T\,x_{2,k}/L \\ \\ T\,x_{1,k}/C \end{bmatrix},$$

then controller (4.66) can be used to achieve trajectory tracking.

4.3.1.2 Control Synthesis

For the case of the boost converter the aim is to control the voltage, which is done by choosing the output as

$$y = x_{2,k}. \tag{4.70}$$

However, for this output selection, the average model (4.68) becomes a nonminimum phase system. It is known that exact trajectory tracking for y as in (4.70) cannot be achieved in a nonminimum phase system [39, 44]. Any control technique for accomplishing exact tracking would render the closed-loop system unstable. A basic way to solve this problem due to the nonminimum phase characteristic is to control the voltage indirectly by means of controlling the inductor current; then the output to be controlled is the inductor current variable defined as

$$y = x_{1,k}. \tag{4.71}$$

Selecting the output system to be (4.71), the system becomes minimum phase and different control techniques for trajectory tracking can be used. However, it is necessary to find the inductor current reference to indirectly controlling the voltage in the capacitor. If the output voltage is constant, the stationary relation between the output voltage and the inductor current is algebraic. Such a relation can be determined from the system equilibrium point, which takes the following form:

$$\bar{x}_{1\delta} = \frac{v_{in}}{\bar{u}^2 R_L}, \qquad \bar{x}_{2\delta} = \frac{v_{in}}{\bar{u}} \tag{4.72}$$

where $\bar{u}$, $\bar{x}_{1\delta}$ and $\bar{x}_{2\delta}$ represent steady-state values. Solving the second term in (4.72) for $\bar{u}$ and replacing $\bar{u}$ in the first one results in

$$\bar{x}_{1\delta} = \frac{\bar{x}_{2\delta}^2}{R_L v_{in}}. \tag{4.73}$$

Relation (4.73) has been extensively employed to control the boost DC-to-DC converter through the inductor current [30]. Hence, reference signals for controller (4.66), to achieve trajectory tracking for system (4.69), become

$$x_{\delta,k} = \begin{bmatrix} x_{1\delta,k} \\ x_{2\delta,k} \end{bmatrix} = \begin{bmatrix} \bar{x}_{1\delta} \\ \bar{x}_{2\delta} \end{bmatrix}.$$

4.3.1.3 Simulation Results

The parameters used for simulation purposes are $L = 12$ mH, $R_L = 220\ \Omega$, $C = 15\ \mu$F, and $v_{in} = 48$ V. The sampling time is selected as $T = 33.33 \times 10^{-6}$ s. The parameters for the inverse optimal control law (4.66), in order to achieve trajectory tracking for the boost converter, are selected as $P = \begin{bmatrix} 2 & 1.5 \\ 1.5 & 2 \end{bmatrix}$ and $R = 0.5$.

Figure 4.7 presents the trajectory tracking time response by considering a constant reference for $x_{2,k} = 100$ V. The respective value for $x_{1,\delta}$ is calculated from (4.73). The

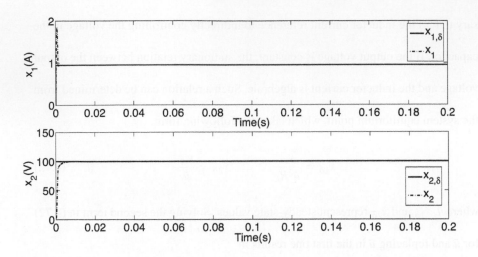

FIGURE 4.7   Trajectory tracking for a constant reference.

initial conditions for the converter are $[x_{1,0}, x_{2,0}]^T = [0, 0]^T$. Figure 4.8 displays the

applied inverse optimal control signal (4.66), which corresponds to the duty cycle.

Figure 4.9 presents the trajectory tracking time response for a time-variant reference

of x_k, with initial conditions $[x_{1,0}, x_{2,0}]^T = [0, 0]^T$. The trajectory to be tracked is

$x_{2,\delta} = 100 + 30\sqrt{2}\sin(20\pi t)$ V and the respective value for $x_{1,\delta}$ is calculated from

(4.73). Figure 4.10 depicts the applied control signal (4.66), which corresponds to the

duty cycle to achieve trajectory tracking of a time-variant reference.

4.4 CLF-BASED INVERSE OPTIMAL CONTROL FOR A CLASS OF

NONLINEAR POSITIVE SYSTEMS

In Section 3.3, a comprehensive explanation of positive nonlinear systems is given.

The trajectory tracking for this class of nonlinear systems is presented in this section

based on the CLF approach as follows.

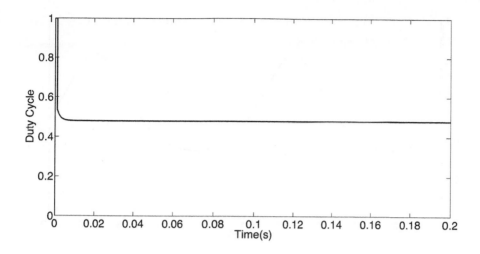

FIGURE 4.8   Control law to track a constant reference.

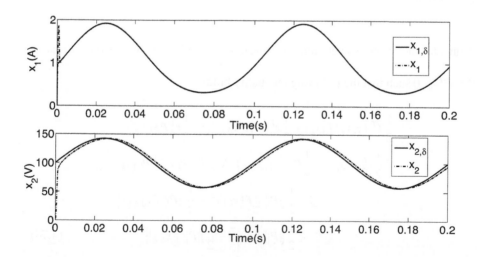

FIGURE 4.9   Trajectory tracking for a time-variant reference.

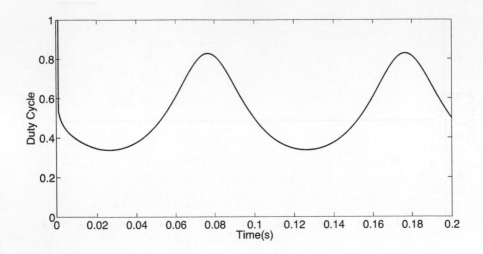

FIGURE 4.10   Control signal to track a time-variant reference.

Theorem 4.3

Consider the affine discrete-time nonlinear system (2.1). If there exists a matrix $P = P^T > 0$ such that the following inequality holds:

$$\frac{1}{2} f^T (x_k) P f(x_k) + \frac{1}{2} x_{\delta,k+1}^T P x_{\delta,k+1} - \frac{1}{2} x_k^T P x_k$$

$$- \frac{1}{2} x_{\delta,k}^T P x_{\delta,k} - \frac{1}{4} P_1^T (x_k, x_{\delta,k}) (R + P_2 (x_k))^{-1} P_1 (x_k, x_{\delta,k})$$

$$\leq -\frac{1}{2} \|P\| \, \|f(x_k)\|^2 - \frac{1}{2} \|P\| \, \|x_{\delta,k+1}\|^2$$

$$- \frac{1}{2} \|P\| \, \|x_k\|^2 - \frac{1}{2} \|P\| \, \|x_{\delta,k}\|^2 \qquad (4.74)$$

where $P_1 (x_k, x_{\delta,k})$ and $P_2 (x_k)$ are defined as

$$P_1 (x_k, x_{\delta,k}) = \begin{cases} g^T (x_k) P \left(f(x_k) - x_{\delta,k+1} \right) & \text{for } f(x_k) \succeq x_{\delta,k+1} \\ g^T (x_k) P \left(x_{\delta,k+1} - f(x_k) \right) & \text{for } f(x_k) \preceq x_{\delta,k+1} \end{cases} \qquad (4.75)$$

and

$$P_2(x_k) = \frac{1}{2} g^T(x_k) P g(x_k) \tag{4.76}$$

respectively, then system (2.1) with control law

$$
\begin{aligned}
u_k^* &= \left| -\frac{1}{4} R^{-1} g^T(x_k) \frac{\partial z_{k+1}^T P z_{k+1}}{\partial z_{k+1}} \right| \\
&= \left| -\frac{1}{2} (R + P_2(x_k))^{-1} P_1(x_k, x_{\delta,k}) \right| \tag{4.77}
\end{aligned}
$$

guarantees asymptotic trajectory tracking along the desired trajectory $x_{\delta,k}$, where

$z_{k+1} = x_{k+1} - x_{\delta,k+1}.$ ■

PROOF

Case 1. Consider the first case for $P_1(x_k, x_{\delta,k})$ in (4.75), that is, $P_1(x_k, x_{\delta,k}) = g^T(x_k) P(f(x_k) - x_{\delta,k+1})$. System (2.1) with control law (4.77) and (4.64), must satisfy inequality (4.63). Considering one step ahead for z_k, we have

$$
\begin{aligned}
\overline{V} &= \frac{1}{2} z_{k+1}^T P z_{k+1} - \frac{1}{2} z_k^T P z_k + u_k^{*T} R u_k^* \\
&= \frac{1}{2} (x_{k+1} - x_{\delta,k+1})^T P (x_{k+1} - x_{\delta,k+1}) - \frac{1}{2} (x_k - x_{\delta,k})^T P (x_k - x_{\delta,k}) \\
&\quad + u_k^{*T} R u_k^*. \tag{4.78}
\end{aligned}
$$

Substituting (2.1) and (4.77) in (4.78), then

$$
\begin{aligned}
\overline{V} &= \frac{1}{2}\left(f\left(x_k\right)+g\left(x_k\right)u_k^*-x_{\delta,k+1}\right)^T P\left(f\left(x_k\right)+g\left(x_k\right)u_k^*-x_{\delta,k+1}\right) \\
&\quad -\frac{1}{2}\left(x-x_{\delta,k}\right)^T P\left(x_k-x_{\delta,k}\right)+u_k^{*T}Ru_k^* \\
&= \frac{1}{2}f^T\left(x_k\right)Pf\left(x_k\right)+\frac{1}{2}u_k^{*T}g^T\left(x_k\right)Pg\left(x_k\right)u_k^*+\frac{1}{2}x_{\delta,k+1}^T Px_{\delta,k+1} \\
&\quad +\frac{1}{2}f^T\left(x_k\right)Pg\left(x_k\right)u_k^*+\frac{1}{2}u_k^{*T}g^T\left(x_k\right)Pf\left(x_k\right)-\frac{1}{2}f^T\left(x_k\right)Px_{\delta,k+1} \\
&\quad -\frac{1}{2}x_{\delta,k+1}^T Pf^T\left(x_k\right)-\frac{1}{2}u_k^T g^T\left(x_k\right)Px_{\delta,k+1}-\frac{1}{2}x_{\delta,k+1}^T Pg\left(x_k\right)u_k^* \\
&\quad -\frac{1}{2}x_k^T Px_k-\frac{1}{2}x_{\delta,k}^T Px_{\delta,k}+\frac{1}{2}x_{\delta,k}^T Px_k+\frac{1}{2}x_k^T Px_{\delta,k}+u_k^{*T}Ru_k^* \quad\quad (4.79)
\end{aligned}
$$

By simplifying, (4.79) becomes

$$
\begin{aligned}
\overline{V} &= \frac{1}{2}f^T\left(x_k\right)Pf\left(x_k\right)+\frac{1}{2}u_k^{*T}g^T\left(x_k\right)Pg\left(x_k\right)u+\frac{1}{2}x_{\delta,k+1}^T Px_{\delta,k+1} \\
&\quad +f^T\left(x_k\right)Pg\left(x_k\right)u_k^*-f^T\left(x_k\right)Px_{\delta,k+1}-x_{\delta,k+1}^T Pg\left(x_k\right)u_k^* \\
&\quad -\frac{1}{2}x_k^T Px_k-\frac{1}{2}x_{\delta,k}^T Px_{\delta,k}+x_k^T Px_{\delta,k}+u_k^{*T}Ru_k^* \\
&= \frac{1}{2}f^T\left(x_k\right)Pf\left(x_k\right)+\frac{1}{2}x_{\delta,k+1}^T Px_{\delta,k+1}+x_{\delta,k+1}^T Pg\left(x_k\right)u_k \\
&\quad -x_{\delta,k+1}^T Pg\left(x_k\right)u_k^*-f^T\left(x_k\right)Px_{\delta,k+1}+x_k^T Px_{\delta,k}-\frac{1}{2}x_k^T Px_k \\
&\quad -\frac{1}{2}x_{\delta,k}^T Px_{\delta,k}+P_1^T\left(x_k,x_{\delta,k}\right)u_k^*+u_k^{*T}P_2\left(x_k\right)u_k^*+u_k^{*T}Ru_k^* \quad\quad (4.80)
\end{aligned}
$$

which after using he control law (4.77), (4.80) results in

$$
\begin{aligned}
\overline{V} &= \frac{1}{2}f^T\left(x_k\right)Pf\left(x_k\right)+\frac{1}{2}x_{\delta,k+1}^T Px_{\delta,k+1}-\frac{1}{2}x_k^T Px_k-\frac{1}{2}x_{\delta,k}^T Px_{\delta,k} \\
&\quad -\frac{1}{4}P_1^T\left(x_k,x_{\delta,k}\right)\left(R+P_2\left(x_k\right)\right)^{-1}P_1\left(x_k,x_{\delta,k}\right) \\
&\quad -f^T\left(x_k\right)Px_{\delta,k+1}+x_k^T Px_{\delta,k} \quad\quad (4.81)
\end{aligned}
$$

Analyzing the sixth and seventh right-hand side (RHS) terms of (4.81) by using the

inequality $X^T Y+Y^T X \leq X^T \Lambda X+Y^T \Lambda^{-1}Y$ proved in [118], which is valid for any vector $X \in \mathbb{R}^{n \times 1}$, then for the sixth RHS term of (4.81), we have

$$
\begin{aligned}
f^T(x_k)Px_{\delta,k+1} &\leq \frac{1}{2}\left[f^T(x_k)Pf(x_k) + (Px_{\delta,k+1})^T P^{-1}(Px_{\delta,k+1})\right] \\
&\leq \frac{1}{2}\left[f^T(x_k)Pf(x_k) + x_{\delta,k+1}^T Px_{\delta,k+1}\right] \\
&\leq \frac{1}{2}\|P\|\|f\|^2 + \frac{1}{2}\|P\|\|x_{\delta,k+1}\|^2.
\end{aligned}
\tag{4.82}
$$

The seventh RHS term of (4.81) becomes

$$
\begin{aligned}
x_k^T Px_{\delta,k} &\leq \frac{1}{2}\left[x_k^T Px_k + (Px_{\delta,k})^T P^{-1}(Px_{\delta,k})\right] \\
&\leq \frac{1}{2}\left[x_k^T Px_k + (Px_{\delta,k})^T P(x_{\delta,k})\right].
\end{aligned}
\tag{4.83}
$$

From (4.83), the following expression holds.

$$
\frac{1}{2}\left[x_k^T Px_k + (Px_{\delta,k})^T P^{-1}(Px_{\delta,k})\right] \leq \frac{1}{2}\|P\|\|x_k\|^2 + \frac{1}{2}\|P\|\|x_{\delta,k}\|.
\tag{4.84}
$$

Substituting (4.82) and (4.84) into (4.81), then

$$
\begin{aligned}
\overline{V} &= \frac{1}{2}f^T(x_k)Pf(x_k) + \frac{1}{2}x_{\delta,k+1}^T Px_{\delta,k+1} - \frac{1}{2}x_k^T Px_k \\
&\quad - \frac{1}{2}x_{\delta,k}^T Px_{\delta,k} - \frac{1}{4}P_1^T(x_k, x_{\delta,k})(R + P_2(x_k))^{-1} P_1(x_k, x_{\delta,k}) \\
&\quad + \frac{1}{2}\|P\|\|f(x_k)\|^2 + \frac{1}{2}\|P\|\|x_{\delta,k+1}\|^2 \\
&\quad + \frac{1}{2}\|P\|\|x_k\|^2 + \frac{1}{2}\|P\|\|x_{\delta,k}\|^2
\end{aligned}
\tag{4.85}
$$

In order to achieve asymptotic stability, it is required that $\overline{V} \leq 0$, then from (4.85) the next inequality is formulated

$$\frac{1}{2}f^T(x_k)Pf(x_k) + \frac{1}{2}x_{\delta,k+1}^T Px_{\delta,k+1} - \frac{1}{2}x_k^T Px_k$$

$$- \frac{1}{2}x_{\delta,k}^T Px_{\delta,k} - \frac{1}{4}P_1^T(x_k,x_{\delta,k})(R+P_2(x_k))^{-1}P_1(x_k,x_{\delta,k})$$

$$\leq -\frac{1}{2}\|P\|\,\|f(x_k)\|^2 - \frac{1}{2}\|P\|\,\|x_{\delta,k+1}\|^2$$

$$-\frac{1}{2}\|P\|\,\|x_k\|^2 - \frac{1}{2}\|P\|\,\|x_{\delta,k}\|^2 \tag{4.86}$$

Hence, selecting P such that (4.86) is satisfied, system (2.1) with control law (4.77) guarantees asymptotic trajectory tracking along the desired trajectory $x_{\delta,k}$. It is worth noting that P and R are positive definite and symmetric matrices; thus, the existence of the inverse in (4.77) is ensured.

Case 2. $P_1(x_k,x_{\delta,k}) = g^T(x_k)P\left(x_{\delta,k+1} - f(x_k)\right)$. It can be derived in the same way as in case 1.

Finally the proposed inverse optimal control law is given as

$$u_k^* = \left| -\frac{1}{2}(R+P_2(x_k))^{-1}P_1(x_k,x_{\delta,k}) \right| \tag{4.87}$$

which ensures that $P_1(x_k,x_{\delta,k})$ satisfies (4.75).

Inverse optimality follows closely that given in Theorem 4.1 and hence it is omitted.

∎

Applicability of the inverse optimal control scheme based on CLF for positive systems is illustrated for glucose control in Chapter 7.

4.5   CONCLUSIONS

This chapter has established the inverse optimal control technique for a class of discrete-time nonlinear systems. To avoid the solution of the HJB equation, we propose a discrete-time CLF in a quadratic form, which depends on a fixed parameter to achieve stabilization and trajectory tracking. Based on this CLF, the inverse optimal control strategy is synthesized. Simulation results illustrate that the proposed controller ensures stabilization and trajectory tracking for nonlinear systems, and a cost functional is minimized. Finally, the inverse optimal control via CLF is used to achieve trajectory tracking for a class of positive nonlinear systems.

5 Inverse Optimal Control:

A CLF Approach, Part II

In this chapter, the inverse optimal control based on a CLF and the discrete-time speed-gradient algorithm is proposed. Section 5.1 uses the speed-gradient algorithm to compute the CLF parameter, ensuring stability and optimality. These results are extended for the speed-gradient inverse optimal control to achieve trajectory tracking in Section 5.2. Additionally, in Section 5.3 an inverse optimal trajectory tracking for block control form nonlinear systems is proposed. Simulation results illustrate the applicability of the proposed control schemes.

5.1 SPEED-GRADIENT ALGORITHM FOR THE INVERSE OPTIMAL CONTROL

In Section 4.1, a CLF candidate such as $V(x_k) = \frac{1}{2}x_k^T P x_k$ is proposed in order to solve the inverse optimal control as established in Definition 4.1, for which an adequate selection of the fixed matrix P must be done such that condition (4.7) is fulfilled. In this section, we propose using the speed-gradient algorithm to calculate this matrix P in a recursive way to ensure the fulfillment of condition (4.7). Then, a CLF candidate

$V(x_k)$ described by

$$V(x_k) = \frac{1}{2} x_k^T P_k x_k, \qquad P_k = P_k^T > 0 \qquad (5.1)$$

is proposed for control law (4.1) in order to guarantee stability for the equilibrium point $x_k = 0$ of system (2.1). Stability will be achieved by defining an appropriate matrix P_k. Moreover, it will be established that control law (4.1) based on (5.1) optimizes the cost functional (2.2).

Consequently, by considering $V(x_k) = V_c(x_k)$ as in (5.1), the control law (4.1) takes the following form:

$$u_k^* = -\frac{1}{2} \left(R + \frac{1}{2} g^T(x_k) P_k g(x_k) \right)^{-1} g^T(x_k) P_k f(x_k). \qquad (5.2)$$

It is worth pointing out that P_k and R are positive definite matrices; thus, the existence of the inverse in (5.2) is assured.

To determine P_k, which ensures stability of the equilibrium point $x_k = 0$ of system (2.1) with (5.2), in this section we propose using the speed-gradient (SG) algorithm.

5.1.1 SPEED-GRADIENT ALGORITHM

The goal of the discrete-time SG algorithm, proposed in [33], is to determine a parameter p, which ensures the following goal:

$$\mathcal{Q}(p) \leq \Delta, \qquad \text{for } k \geq k^*, \qquad (5.3)$$

where $\mathcal{Q}$ is a positive definite goal function, Δ is a positive constant considered as a threshold, and $k^* \in \mathbb{Z}^+$ is the time step at which the goal is achieved. Related results on the continuous-time and discrete-time speed-gradient algorithms and their

applications are given in [33, 88] and references therein.

Control law (5.2) at every time step depends on the matrix P_k, which is defined as

$$P_k = p_k P'$$

where $P' = P'^T > 0$ is a given constant matrix and p_k is a scalar parameter to be

determined by the SG algorithm. Then, (5.2) is transformed into

$$u_k^* = -\frac{p_k}{2} \left(R + \frac{p_k}{2} g^T(x_k) P' g(x_k) \right)^{-1} g^T(x_k) P' f(x_k). \tag{5.4}$$

The SG algorithm is now reformulated for the inverse optimal control.

DEFINITION 5.1: SG Goal Function Consider a time-varying parameter $p_k \in \mathbb{R}^+$.

The positive definite $\mathscr{C}^1$ function $\mathscr{Q} : \mathbb{R}^n \times \mathbb{R}^+ \to \mathbb{R}^+$ given as

$$\mathscr{Q}(x_k, p_k) = V_{sg}(x_{k+1}) \tag{5.5}$$

where $V_{sg}(x_{k+1}) = \frac{1}{2} x_{k+1}^T P' x_{k+1}$, with $x_{k+1} = f(x_k) + g(x_k) u_k^*$, is named as the SG

goal function for system (2.1) with control law (5.4).

The SG goal function is defined as in (5.5) in such a way that the convexity property

of $\mathscr{Q}(x_k, p_k)$ for p_k is guaranteed; then there exist an optimal value p^* for p_k and a

positive constant ε^* such that $\mathscr{Q}(x_k, p^*) \leq \varepsilon^*$ [33]. In Theorem 5.1 below, this SG

goal function is used to construct a Lyapunov function for the closed-loop system.

DEFINITION 5.2: SG Control Goal The SG control goal for system (2.1) with

(5.4) is defined as

$$\mathcal{Q}(x_k, p_k) \leq \Delta(x_k), \qquad \text{for} \quad k \geq k^* \tag{5.6}$$

where p_k is the value such that (5.6) is fulfilled, with

$$\Delta(x_k) = V_{sg}(x_k) - \frac{1}{p_k} u_k^{*T} R u_k^*, \tag{5.7}$$

$V_{sg}(x_k) = \frac{1}{2} x_k^T P' x_k$ and $k^* \in \mathbb{Z}_+$ is the time step at which the SG control goal is

achieved.

REMARK 5.1 Solution p_k must guarantee that $V_{sg}(x_k) > \frac{1}{p_k} u_k^{*T} R u_k^*$ in order to

obtain a positive definite function $\Delta(x_k)$. ∎

Let us state the first contribution of the chapter as follows.

Lemma 5.1

Consider the discrete-time nonlinear system (2.1) with (5.4) as input. Let $\mathcal{Q}$ be an

SG goal function as defined in (5.5). Let $\bar{p}$ be a positive constant, $\Delta(x_k)$ be a positive

definite function with $\Delta(0) = 0$, and assume that there exist positive constants p^* and

ε^* such that the following control goal is achievable [33]:

$$\mathcal{Q}(x_k, p^*) \leq \varepsilon^* \ll \Delta(x_k) \tag{5.8}$$

Then, for any initial condition $p_0 > 0$, there exists a $k^* \in \mathbb{Z}^+$ such that the SG control goal (5.6) is achieved by means of the following dynamic variation of parameter p_k:

$$p_{k+1} = p_k - \gamma_{d,k} \nabla_p \mathcal{Q}(x_k, p_k), \qquad (5.9)$$

with

$$\gamma_{d,k} = \gamma_c \, \delta_k \left| \nabla_p \mathcal{Q}(x_k, p_k) \right|^{-2}, \qquad 0 < \gamma_c \leq 2\Delta(x_k)$$

and

$$\delta_k = \begin{cases} 1 & for \quad \mathcal{Q}(x_k, p_k) > \Delta(x_k) \\ \\ 0 & otherwise. \end{cases} \qquad (5.10)$$

Finally, for $k \geq k^*$, p_k becomes a constant denoted by $\bar{p}$ and the SG algorithm terminates.   ∎

PROOF

Along the lines of [33], the proof is based on the case for which $\mathcal{Q}(x_k, p_k) > \Delta(x_k)$, and therefore $\delta_k = 1$. Let us consider the positive definite Lyapunov function $V_p(p_k) = |p_k - p^*|^2$. Then, the respective Lyapunov difference is given as

$$\begin{aligned} \Delta V_p(p_k) &= |p_{k+1} - p^*|^2 - |p_k - p^*|^2 \\ &= (p_{k+1} - p_k)^T [(p_{k+1} - p_k) + 2(p_k - p^*)] \\ &= -\gamma_{d,k} \nabla_p \mathcal{Q}(x_k, p_k) [-\gamma_{d,k} \nabla_p \mathcal{Q}(x_k, p_k) + 2(p_k - p^*)] \quad (5.11) \end{aligned}$$

Due to convexity of the SG goal function (5.5) for p_k,

$$(p^* - p_k)^T \nabla_p \mathcal{Q}(x_k, p_k) \leq \varepsilon^* - \Delta(x_k) < 0 \qquad (5.12)$$

where $\nabla_p \mathscr{Q}(x_k, p_k)$ denotes the gradient of $\mathscr{Q}(x_k, p_k)$ with respect to p_k. Based on

(5.12), (5.11) becomes

$$
\begin{aligned}
\Delta V_p(p_k) &\leq -2\gamma_{d,k}\left(\Delta(x_k) - \varepsilon^*\right) + \gamma_{d,k}^2 \left|\nabla_p \mathscr{Q}(x_k, p_k)\right|^2 \\
&\leq -2\gamma_c \delta_k \left(\Delta(x_k) - \varepsilon^*\right)\left|\nabla_p \mathscr{Q}(x_k, p_k)\right|^{-2} \\
&\quad + \gamma_c^2 \delta_k^2 \left|\nabla_p \mathscr{Q}(x_k, p_k)\right|^{-4}\left|\nabla_p \mathscr{Q}(x_k, p_k)\right|^2 \\
&= -\frac{\gamma_c \left[2\Delta(x_k)\left(1 - \left(\varepsilon^*/\Delta(x_k)\right)\right) - \gamma_c\right]}{\left|\nabla_p \mathscr{Q}(x_k, p_k)\right|^2}.
\end{aligned}
$$

From (5.8), $1 - \left(\varepsilon^*/\Delta(x_k)\right) \approx 1$, hence

$$
\begin{aligned}
\Delta V_p(p_k) &\approx -\frac{\gamma_c \left[2\Delta(x_k) - \gamma_c\right]}{\left|\nabla_p \mathscr{Q}(x_k, p_k)\right|^2} \\
&< 0.
\end{aligned}
$$

Thus, boundedness of p_k is guaranteed if $0 < \gamma_c \leq 2\Delta(x_k)$. Finally, when $k \geq k^*$, then

$\delta_k = 0$, which means the algorithm terminates; at this point $\mathscr{Q}(x_k, p_k) \leq \Delta(x_k)$, then

p_k becomes a constant value denoted by $\bar{p}$ (i.e., $p_k = \bar{p}$).   ■

Note that the gradient $\nabla_p \mathscr{Q}(x_k, p_k)$ in (5.9) is reduced to being the partial derivative

of $\mathscr{Q}(x_k, p_k)$ with respect to p_k, as $\frac{\partial}{\partial p_k} \mathscr{Q}(x_k, p_k)$.

REMARK 5.2 Parameter γ_c in (5.9) is selected such that the solution p_k ensures the

requirement $V_{sg}(x_k) > \frac{1}{p_k} u_k^T R u_k$ in Remark 5.1. Then, we have a positive definite

function $\Delta(x_k)$.   ■

REMARK 5.3 With $\mathscr{Q}(x_k, p_k)$ as defined in (5.5), the dynamic variation of parameter p_k in (5.9) results in

$$p_{k+1} = p_k + 8\gamma_{d,k} \frac{f^T(x_k) P' g(x_k) R^2 g^T(x_k) f(x_k)}{\left(2R + p_k g^T(x_k) P' g(x_k)\right)^3}$$

which is positive for all time steps k if $p_0 > 0$. Therefore positiveness for p_k is ensured and the requirement $P_k = P_k^T > 0$ for (5.1) is guaranteed. ∎

When the SG control goal (5.6) is achieved, then $p_k = \bar{p}$ for $k \geq k^*$. Thus, matrix P_k in (5.2) is considered constant and $P_k = P$, where P is computed as $P = \bar{p}P'$, with P' a design positive definite matrix. Under these constraints, we obtain

$$
\begin{aligned}
\alpha(x_k) \quad &:= \quad u_k^* \\
&= \quad -\frac{1}{2}\left(R + \frac{1}{2}g^T(x_k) P g(x_k)\right)^{-1} g^T(x_k) P f(x_k).
\end{aligned}
\tag{5.13}
$$

Figure 5.1 presents the flow diagram for the proposed SG algorithm.

5.1.2 SUMMARY OF THE PROPOSED SG ALGORITHM TO CALCULATE PARAMETER P_K

Considering the closed-loop system (2.1) with (5.4) as input, we obtain

$$x_{k+1} = f(x_k) - \frac{p_k}{2}g(x_k)\left(R + \frac{p_k}{2}g^T(x_k) P' g(x_k)\right)^{-1} g^T(x_k) P' f(x_k).$$

Then, we propose the SG goal function

$$\mathscr{Q}_k(p_k) = x_{k+1}^T x_{k+1}.$$

The dynamic variation of parameter p_k is established as

$$p_{k+1} = p_k - \gamma \nabla_p \mathscr{Q}_k(p_k), \qquad p_0 = p(0).$$

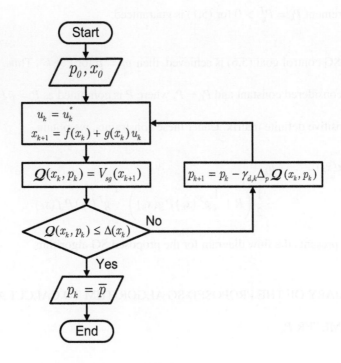

FIGURE 5.1   Speed-gradient algorithm flow diagram.

Finally, when condition (5.6) is fulfilled, the SG algorithm finishes.

5.1.3 SG INVERSE OPTIMAL CONTROL

Once the control law (5.13) has been determined, we proceed to demonstrate that it

ensures stability and optimality for (2.1) without solving the HJB equation (2.10).

Thus, the second contribution of this chapter is stated as the following theorem.

Theorem 5.1

Consider that system (2.1) with (5.4) has achieved the SG control goal (5.6) by

means of (5.9). Let $V(x_k) = \frac{1}{2}x_k^T P x_k$ be a Lyapunov function candidate with $P =$

$P^T > 0$. Then, control law (5.13) is an inverse optimal control law, in accordance

with Definition 4.1, which ensures that the equilibrium point $x_k = 0$ of system (2.1)

is globally asymptotically stable. Moreover, with $V(x_k) = \frac{1}{2}x_k^T P x_k$ as a CLF and

$P = \bar{p}P'$, control law (5.13) is inverse optimal in the sense that it minimizes the cost

functional given by

$$\mathscr{J} = \sum_{k=0}^{\infty} \left(l(x_k) + u_k^T R\, u_k \right) \tag{5.14}$$

where

$$l(x_k) := -\overline{V} \tag{5.15}$$

with $\overline{V}$ defined as

$$\overline{V} = V(x_{k+1}) - V(x_k) + \alpha^T(x_k) R\, \alpha(x_k).$$

■

PROOF

Considering that system (2.1), with control law (5.4) and the SG algorithm (5.9)

achieves the SG control goal (5.6) for $k \geq k^*$ (Lemma 5.1), then (5.6) can be rewritten

as

$$V_{sg}(x_{k+1}) - V_{sg}(x_k) + \frac{1}{\bar{p}} \alpha^T(x_k) R \alpha(x_k)$$

$$= \frac{1}{2} x_{k+1}^T P' x_{k+1} - \frac{1}{2} x_k^T P' x_k + \frac{1}{\bar{p}} \alpha^T(x_k) R \alpha(x_k)$$

$$\leq 0. \tag{5.16}$$

Multiplying (5.16) by the positive constant $\bar{p}$ of Lemma 5.1, and using the SG goal

function as a Lyapunov function given as $V(x_k) = \bar{p} V_{sg}(x_k)$ for the closed-loop system,

we obtain

$$\begin{aligned}
\bar{V} &:= \frac{\bar{p}}{2} x_{k+1}^T P' x_{k+1} - \frac{\bar{p}}{2} x_k^T P' x_k + \alpha^T(x_k) R \alpha(x_k) \\
&= \frac{1}{2} x_{k+1}^T P x_{k+1} - \frac{1}{2} x_k^T P x_k + \alpha^T(x_k) R \alpha(x_k) \\
&= V(x_{k+1}) - V(x_k) + \alpha^T(x_k) R \alpha(x_k) \\
&\leq 0. \tag{5.17}
\end{aligned}$$

From (5.17), obviously $V(x_{k+1}) - V(x_k) < 0$ for all $x_k \neq 0$ with $V(x_k)$ a positive

definite and radially unbounded function, then global asymptotic stability is achieved

in accordance with Theorem 2.1.

When function $-l(x_k)$ is set to be the right-hand side of (5.17), then

$$l(x_k) \quad := \quad -\overline{V} \tag{5.18}$$

$$= \quad -(V(x_{k+1}) - V(x_k)) - \alpha^T(x_k) R \alpha(x_k)$$

$$\geq \quad 0, \qquad \forall x_k \neq 0.$$

Consequently, $V(x_k) = \frac{1}{2} x_k^T P x_k$ as a CLF is a solution of the HJB equation (2.10) for $k \geq k^*$.

In order to obtain the optimal value function for the cost functional (5.14), we proceed as in Theorem 4.1. ∎

As established in [121], to use a CLF for the inverse optimal control approach, the entries of P' in $V(x_k)$ are selected such that $\Delta V(x_k, u_k)$ is negative definite, which considers the flexibility provided by the control term $g(x_k)u_k$. Additionally, once the condition $\Delta V(x_k, u_k) < 0$ is fulfilled, adequate values for the entries of P' are a matter of trial and error such that a good performance is achieved for the system dynamics. These arguments motivated us to explore the use of the SG algorithm for the selection of matrix P'. Progress is being made in determining P' by means of a matrix inequality constraint.

It is worth mentioning that the CLF approach for control synthesis has been applied successfully to systems for which a CLF can be established, such as feedback linearizable, strict feedback, and feed-forward ones [37, 108]. However, systematic techniques for determining CLFs do not exist for general nonlinear systems [108].

In this work, we refer to (2.2) as a cost functional due to the fact that a weighting for both the state and the control input can be established. The weighting of the control

input in (2.2) is selected directly by R, while for the state it is analyzed as follows:

$$
\begin{aligned}
l(x_k) &= -\left[V(x_{k+1}) - V(x_k) + \alpha^T(x_k)R\alpha(x_k)\right] \\
&= -\frac{f^T(x_k)Pf(x_k) + 2f^T(x_k)Pg(x_k)\alpha(x_k)}{2} \\
&\quad + \frac{x_k^T Px_k - \alpha^T(x_k)g^T(x_k)Pg(x_k)\alpha(x_k)}{2} - \alpha^T(x_k)R\alpha(x_k) \\
&= \frac{1}{2}\left(x_k^T Px_k - f^T(x_k)Pf(x_k)\right) + \frac{1}{2}f^T(x_k)Pg(x_k) \\
&\quad \times \left(R + \frac{1}{2}g^T(x_k)Pg(x_k)\right)^{-1}g^T(x_k)Pf(x_k) - \frac{1}{4}f^T(x_k)Pg(x_k) \\
&\quad \times \left(R + \frac{1}{2}g^T(x_k)Pg(x_k)\right)^{-1}g^T(x_k)Pf(x_k) \\
&= \frac{1}{2}\left(x_k^T Px_k - f^T(x_k)Pf(x_k)\right) + \frac{1}{4}f^T(x_k)Pg(x_k) \\
&\quad \times \left(R + \frac{1}{2}g^T(x_k)Pg(x_k)\right)^{-1}g^T(x_k)Pf(x_k)
\end{aligned}
$$

which can be rewritten as

$$
\begin{aligned}
l(x_k) &= \frac{1}{2}f^T(x_k)\left[\frac{1}{2}Pg(x_k)\left(R + \frac{1}{2}g^T(x_k)Pg(x_k)\right)^{-1}g^T(x_k)P - P\right]f(x_k) \\
&\quad + \frac{1}{2}x_k^T Px_k \\
&= \frac{1}{2}\left[\begin{array}{cc} f(x_k) & x_k \end{array}\right]^T \\
&\quad \times \left[\begin{array}{cc} \frac{1}{2}Pg(x_k)\left(R + \frac{1}{2}g^T(x_k)Pg(x_k)\right)^{-1}g^T(x_k)P - P & 0 \\ 0 & P \end{array}\right] \\
&\quad \times \left[\begin{array}{c} f(x_k) \\ x_k \end{array}\right].
\end{aligned}
$$

Hence, selecting an adequate P for $l(x_k)$, a weight term for x_k can be obtained.

5.1.3.1 Example

We synthesize an inverse optimal control law for a discrete-time second order non-linear system (unstable for $u_k = 0$) of the form (2.1) with

$$f(x_k) = \begin{bmatrix} x_{1,k}^2 x_{2,k} - 0.8 x_{2,k} \\ x_{1,k}^2 + 1.8 x_{2,k} \end{bmatrix} \tag{5.19}$$

and

$$g(x_k) = \begin{bmatrix} 0 \\ -2 + \cos(x_{2,k}) \end{bmatrix}. \tag{5.20}$$

According to (5.13), the inverse optimal control law is formulated as

$$u_k^* = -\frac{1}{2} \left(R + \frac{1}{2} g^T(x_k) P_k g(x_k) \right)^{-1} g^T(x_k) P_k f(x_k)$$

where the positive definite matrix $P_k = p_k P'$ is calculated by the SG algorithm, with P' as the identity matrix, that is,

$$\begin{aligned} P_k &= p_k P' \\ &= p_k \begin{bmatrix} 1 & 0 \\ 0 & 1 \end{bmatrix} \end{aligned}$$

and R is a constant term selected as

$$R = 0.2.$$

For this simulation, the selection of the identity matrix for P' is sufficient to ensure asymptotic stability. The state penalty term $l(x_k)$ in (5.14) is calculated according to (5.15). The phase portrait for this unstable open-loop ($u_k = 0$) system with initial

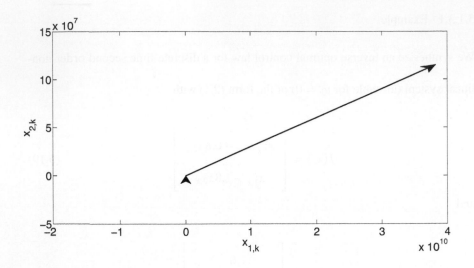

FIGURE 5.2   Open-loop unstable phase portrait.

conditions $\chi_0 = [2 \quad -2]^T$ is displayed in Figure 5.2. Figure 5.3 shows the time

evolution of x_k for this system with initial conditions $x_0 = [2 \ -2]^T$ under the action of

the proposed control law. This figure also includes the applied inverse optimal control

law, which achieves asymptotic stability; the respective phase portrait is displayed in

Figure 5.4. Figure 5.5 displays the SG algorithm solution p_k; the evaluation of the

cost functional $\mathscr{J}$ is also shown in this figure.

5.1.4   APPLICATION TO THE INVERTED PENDULUM ON A CART

The proposed inverse optimal control is illustrated by stabilizing the inverted pendu-

lum on a cart at the upright position [82] (see Figure 5.7), which is difficult to control

due to the fact that is an underactuated system with $\overrightarrow{F}$ the only control input. The

control scheme for the pendulum on a cart could be used in applications such as the

Segway personal transporter (see Figure 5.6).

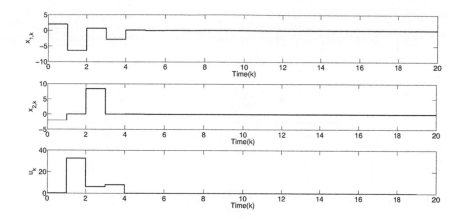

FIGURE 5.3   Stabilization of a nonlinear system using the speed-gradient algorithm.

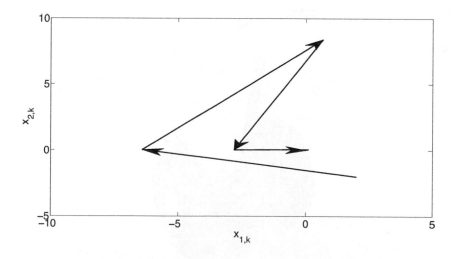

FIGURE 5.4   Phase portrait for the stabilized system using the speed-gradient algorithm.

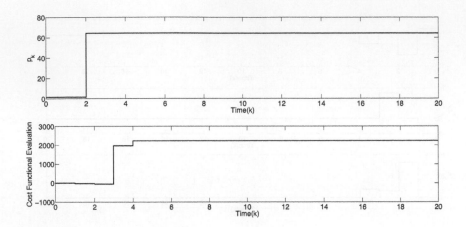

FIGURE 5.5 p_k and $\mathscr{J}$ time evolution.

FIGURE 5.6 (**SEE COLOR INSERT**)   Segway personal transporter.

The dynamics of the inverted pendulum are given as [82]

$$\dot{x} = v_x$$

$$\dot{v}_x = \frac{ml\,\omega^2\sin\theta - mg\sin\theta\cos\theta + \vec{F}}{M + m\sin^2\theta}$$

$$\dot{\theta} = \omega \tag{5.21}$$

$$\dot{\omega} = \frac{-ml\,\omega^2\sin\theta\cos\theta + (M+m)\,g\sin\theta - \vec{F}\cos\theta}{Ml + ml\sin^2\theta}$$

where x is the car position, v_x is the car velocity, θ is the pendulum angle, ω is the angular velocity, M is the mass of the car, m is the point mass attached at the end of the pendulum, l is the length of the pendulum, g is the gravity constant, and $\vec{F}$ is force applied to the cart.

After discretizing by Euler approximation,[1] the discrete-time model for the inverted pendulum on a cart is rewritten as

$$x_{k+1} = x_k + T v_{x,k}$$

$$v_{x,k+1} = v_{x,k} + T\left(\frac{ml\,\omega_k^2\sin\theta_k - mg\sin\theta_k\cos\theta_k}{M + m\sin^2\theta_k}\right) + \frac{T}{M + m\sin^2\theta_k}\vec{F}_k$$

$$\theta_{k+1} = \theta_k + T\,\omega_k$$

$$\omega_{k+1} = \omega_k + T\left(\frac{-ml\,\omega_k^2\sin\theta_k\cos\theta_k + (M+m)g\sin\theta_k}{Ml + ml\sin^2\theta_k}\right)$$
$$+ \frac{-T\cos\theta_k}{Ml + ml\sin^2\theta_k}\vec{F}_k \tag{5.22}$$

where T is the sampling time.

System (5.22) can be presented in a general affine form as

$$x_{k+1} = f(x_k) + g(x_k)\vec{F}_k \tag{5.23}$$

---

[1]For the ordinary differential equation $\frac{dz}{dt} = f(z)$, the Euler discretization is defined as $\frac{z_{k+1}-z_k}{T} = f(z_k)$, such that $z_{k+1} = z_k + T f(z_k)$, where T is the sampling time [67, 74].

where $x_k = [x_k, v_{x,k}, \theta_k, \omega_k]^T$, and the inverse optimal control law (5.13) is applied

for this system as $\overrightarrow{F}_k = \alpha(x_k)$.

5.1.4.1 Simulation Results

The parameters used for simulation are $M = 3\,\text{kg}$, $m = 1\,\text{kg}$, $l = 0.5\,\text{m}$, $g = 9.81\,\text{m/s}^2$,

and the sampling time is $T = 0.001$ s. The initial conditions are $[x_0, v_{x,0}, \theta_0, \omega_0]^T = [0, 0, 0.5, 0]^T$.

For this simulation, the selection of the matrix P' in (5.13) is done as

$$P' = \begin{bmatrix} 2.81 & 1.80 & 1.70 & 1.20 \\ 1.80 & 2.81 & 2.80 & 2.79 \\ 1.70 & 2.80 & 3.00 & 2.90 \\ 1.20 & 2.79 & 2.90 & 3.00 \end{bmatrix}$$

which is sufficient to ensure asymptotic stability on the desired angle position. The

initial condition for p_k in the SG algorithm is $p_0 = 5$. Matrix R in (5.14) is selected

as

$$R = 0.5.$$

Figure 5.8 presents the time evolution of x_k, $v_{x,k}$, θ_k, and ω_k. There, it can be seen

that the stabilization of the inverted pendulum on the upright position ($\theta = 0$ rad) is

achieved. Figure 5.9 displays the applied inverse optimal control law and the evalua-

tion of the respective cost functional.

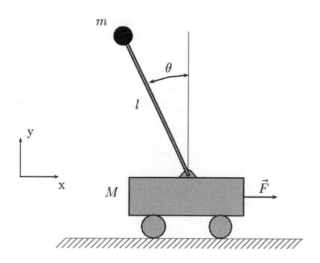

FIGURE 5.7   Inverted pendulum on a cart.

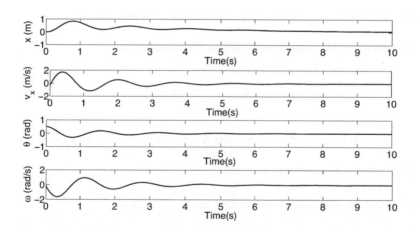

FIGURE 5.8   Stabilized inverted pendulum time response.

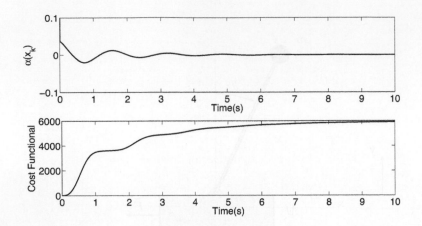

FIGURE 5.9   Control signal and cost functional time evolution for the inverted pendulum.

5.2   SPEED-GRADIENT ALGORITHM FOR TRAJECTORY TRACKING

In this section, instead of determining a fixed matrix P as proposed in (4.64), a time-varying matrix P_k is calculated by using the speed-gradient algorithm, which ensures trajectory tracking of x_k for system (2.1) along the desired trajectory $x_{\delta,k}$.

As in Section 5.1, the control goal function is established as

$$\mathscr{Q}(z_{k+1}) \leq \Delta, \qquad \text{for } k \geq k^* \qquad (5.24)$$

where $\mathscr{Q}$ is a control goal function, a constant $\Delta > 0$, and $k^* \in \mathbb{Z}^+$ is the time at which the control goal is achieved.

Let us define the matrix P_k at every time k as

$$P_k = p_k \overline{P}$$

where p_k is a scalar parameter to be adjusted by the SG algorithm, $\overline{P} = K^T P' K$ with K an additional diagonal gain matrix of appropriate dimension introduced to modify the convergence rate of the tracking error, and $P' = P'^T > 0$ a design constant matrix

of appropriate dimension. Then, in a similar way as established in Section 4.3, the

inverse optimal control law, which uses the SG algorithm, becomes

$$u_k = -\frac{p_k}{2} \left(R + \frac{p_k}{2} g^T(x_k) \overline{P} g(x_k) \right)^{-1} g^T(x_k) \overline{P}(f(x_k) - x_{\delta,k+1}). \qquad (5.25)$$

The SG algorithm is now reformulated for trajectory tracking inverse optimal control.

DEFINITION 5.3: SG Goal Function for Trajectory Tracking Consider a time-

varying parameter $p_k \in \mathscr{P} \subset \mathbb{R}^+$, with $p_k > 0$ for all k, and $\mathscr{P}$ is the set of admissible

values for p_k. A nonnegative $\mathscr{C}^1$ function $\mathscr{Q} : \mathbb{R}^n \times \mathbb{R} \to \mathbb{R}$ of the form

$$\mathscr{Q}(z_k, p_k) = V_{sg}(z_{k+1}), \qquad (5.26)$$

where $V_{sg}(z_{k+1}) = \frac{1}{2} z_{k+1}^T P' z_{k+1}$, is referred to as the SG goal function for system

(2.1), with $z_{k+1} = x_{k+1} - x_{\delta,k+1}$, x_{k+1} as defined in (2.1), control law (5.25), and

desired reference $x_{\delta,k+1}$. We define $\mathscr{Q}_k(p) := \mathscr{Q}(z_k, p_k)$.

DEFINITION 5.4: SG Control Goal for Trajectory Tracking Consider a constant

$p^* \in \mathscr{P}$. The SG control goal for system (2.1) with (5.25) is defined as finding p_k so

that the SG goal function $\mathscr{Q}_k(p)$ as defined in (5.26) fulfills

$$\mathscr{Q}_k(p) \le \Delta(z_k), \qquad \text{for} \quad k \ge k^*, \qquad (5.27)$$

where

$$\Delta(z_k) = V_{sg}(z_k) - \frac{1}{p_k} u_k^T R u_k \qquad (5.28)$$

with $V_{sg}(z_k) = \frac{1}{2} z_k^T P' z_k$ and u_k as defined in (5.25); $k^* \in \mathbb{Z}^+$ is the time at which the

SG control goal is achieved.

REMARK 5.4 Solution p_k must guarantee that $V_{sg}(z_k) > \frac{1}{p_k} u_k^T R u_k$ in order to obtain

a positive definite function $\Delta(z_k)$.   ∎

The SG algorithm is used to compute p_k in order to achieve the SG control goal

defined above.

Lemma 5.2

Consider a discrete-time nonlinear system of the form (2.1) with (5.25) as input. Let

$\mathcal{Q}$ be an SG goal function as defined in (5.26), and denoted by $\mathcal{Q}_k(p)$. Let $\bar{p}, p^* \in \mathcal{P}$

be positive constant values, $\Delta(z_k)$ be a positive definite function with $\Delta(0) = 0$, and

ε^* be a sufficiently small positive constant. Assume that

- A1. There exist p^* and ε^* such that

$$\mathcal{Q}_k(p^*) \leq \varepsilon^* \ll \Delta(z_k) \quad \text{and} \quad 1 - \varepsilon^*/\Delta(z_k) \approx 1. \qquad (5.29)$$

- A2. For all $p_k \in \mathcal{P}$:

$$(p^* - p_k)^T \nabla_p \mathcal{Q}_k(p) \leq \varepsilon^* - \Delta(z_k) < 0 \qquad (5.30)$$

where $\nabla_p \mathcal{Q}_k(p)$ denotes the gradient of $\mathcal{Q}_k(p)$ with respect to p_k.

Then, for any initial condition $p_0 > 0$, there exists a $k^* \in \mathbb{Z}^+$ such that the SG control goal (5.27) is achieved by means of the following dynamic variation of parameter p_k:

$$p_{k+1} = p_k - \gamma_{d,k} \nabla_p \mathcal{Q}_k(p), \tag{5.31}$$

with

$$\gamma_{d,k} = \gamma_c \, \delta_k \left| \nabla_p \mathcal{Q}_k(p) \right|^{-2}, \qquad 0 < \gamma_c \leq 2\Delta(z_k)$$

and

$$\delta_k = \begin{cases} 1 & for \quad Q(p_k) > \Delta(z_k) \\ \\ 0 & otherwise. \end{cases} \tag{5.32}$$

Finally, for $k \geq k^*$, p_k becomes a constant value denoted by $\bar{p}$ and the SG algorithm is completed. ∎

PROOF

It follows closely the proof of Lemma 5.1. ∎

REMARK 5.5 Parameter γ_c in (5.31) is selected such that solution p_k ensures the requirement $V_{sg}(z_k) > \frac{1}{p_k} u_k^T R u_k$ in Remark 5.4. Then, we have a positive definite function $\Delta(z_k)$. ∎

When the SG control goal (5.27) is achieved, then $p_k = \bar{p}$ for $k \geq k^*$. Thus, matrix P_k is considered constant, that is, $P_k = P$, where P is computed as $P = \bar{p} K P' K$, with

P' a *design* positive definite matrix. Under these constraints, we obtain

$$\alpha(z) \quad := \quad u_k$$

$$= \quad -\frac{1}{2}\left(R + \frac{1}{2}g^T(x_k)Pg(x_k)\right)^{-1} g^T(x_k)P(f(x_k) - x_{\delta,k+1}). \quad (5.33)$$

The following theorem establishes the trajectory tracking via inverse optimal control.

Theorem 5.2

Consider that system (2.1) with (5.25) has achieved the SG control goal (5.27) by means of (5.31). Let $V(z_k) = \frac{1}{2}z_k^T P z_k$ be a Lyapunov function candidate with $P = P^T > 0$. Then, trajectory tracking inverse optimal control law (5.33) renders solution x_k of system (2.1) to be globally asymptotically stable along the desired trajectory $x_{\delta,k}$. Moreover, with $V(x_k) = \frac{1}{2}z_k^T P z_k$ as a CLF and $P = \bar{p}P'$, this control law (5.33) is inverse optimal in the sense that it minimizes the cost functional given by

$$\mathscr{J}(z_k) = \sum_{k=0}^{\infty}\left(l(z_k) + u_k^T R\, u_k\right) \quad (5.34)$$

where

$$l(z_k) := -\overline{V} \quad (5.35)$$

with $\overline{V}$ defined as

$$\overline{V} = V(z_{k+1}) - V(z_k) + \alpha^T(z)R\alpha(z)$$

and $\alpha(z)$ as defined in (5.33). ∎

PROOF

It follows closely the proof given for Theorem 5.1 and hence it is omitted. ∎

5.2.1 EXAMPLE

To illustrate the applicability of the proposed methodology, we synthesize a trajectory tracking inverse optimal control law in order to achieve trajectory tracking for a discrete-time second order nonlinear system (unstable for $u_k = 0$) of the form (2.1) with

$$f(x_k) = \begin{bmatrix} 2x_{1,k} \sin(0.5x_{1,k}) + 0.1x_{2,k}^2 \\ 0.1x_{1,k}^2 + 1.8x_{2,k} \end{bmatrix} \tag{5.36}$$

and

$$g(x_k) = \begin{bmatrix} 0 \\ 2 + 0.1\cos(x_{2,k}) \end{bmatrix}. \tag{5.37}$$

According to (5.33), the trajectory tracking inverse optimal control law is formulated as

$$u_k = -\frac{1}{2}\left(R + \frac{1}{2}g^T(x_k)Pg(x_k)\right)^{-1} g^T(x_k)P(f(x_k) - x_{\delta,k+1})$$

where the positive definite matrix $P_k = p_k P'$ is calculated by the SG algorithm with

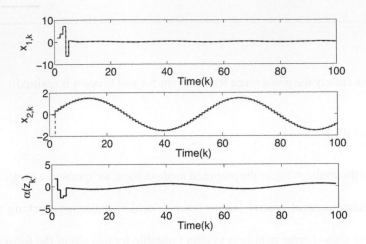

FIGURE 5.10 (**SEE COLOR INSERT**)   Tracking performance of x_k.

P' as the identity matrix, that is,

$$
\begin{aligned}
P_k &= p_k P' \\
&= p_k \begin{bmatrix} 0.020 & 0.016 \\ 0.016 & 0.020 \end{bmatrix}
\end{aligned}
$$

and R is a constant matrix

$$R = 0.5.$$

The reference for $x_{2,k}$ is

$$x_{2\delta,k} = 1.5\,sin(0.12k)\ rad.$$

and reference $x_{1\delta,k}$ is defined accordingly.

Figure 5.10 presents the trajectory tracking for x_k with initial condition $p_0 = 2.5$ for the SG algorithm, where the solid line $(x_{\delta,k})$ is the reference signal and the dashed line is the evolution of x_k. The control signal is also displayed.

5.3 TRAJECTORY TRACKING FOR SYSTEMS IN BLOCK-CONTROL

FORM

In this section, trajectory tracking is established as a stabilization problem based on an

error coordinate transformation for systems in the block-control form. Let us consider

that system (2.1) can be presented (possibly after a nonlinear transformation) in the

nonlinear block-control form [72] consisting of r blocks as

$$x_{k+1}^1 \ = \ f^1\left(x_k^1\right) + B^1\left(x_k^1\right) x_k^2$$

$$\vdots$$

$$x_{k+1}^{r-1} \ = \ f^{r-1}\left(x_k^1, x_k^2, \ldots, x_k^{r-1}\right) \tag{5.38}$$

$$+ B^{r-1}\left(x_k^1, x_k^2, \ldots, x_k^{r-1}\right) x_k^r$$

$$x_{k+1}^r \ = \ f^r\left(x_k\right) + B^r\left(x_k\right) \alpha(x_k)$$

where $x_k \in \mathbb{R}^n$, $x_k = \begin{bmatrix} x_k^{1T} & x_k^{2T} & \cdots & x_k^{rT} \end{bmatrix}^T$; $x^j \in \mathbb{R}^{n_j}$; $j = 1, \ldots, r$; n_j denotes the order

of each r-th block and $n = \sum_{j=1}^r n_j$; input $\alpha(x_k) \in \mathbb{R}^m$; $f^j : \mathbb{R}^{n_1 + \cdots + n_j} \to \mathbb{R}^{n_j}$, $B^j :$

$\mathbb{R}^{n_1 + \cdots + n_j} \to \mathbb{R}^{n_j \times n_{j+1}}$, $j = 1, \ldots, r-1$, and $B^r : \mathbb{R}^n \to \mathbb{R}^{n_r \times m}$ are smooth mappings.

Without loss of generality, $x_k = 0$ is an equilibrium point for (5.38). We assume

$f^j(0) = 0$, $rank\{B^j(x_k)\} = n_j \ \forall x_k \neq 0$.

For trajectory tracking of the first block in (5.38), let us define the tracking error

as

$$z_k^1 = x_k^1 - x_{\delta,k}^1 \tag{5.39}$$

where $x_{\delta,k}^j$ is the desired trajectory.

Once the first new variable is determined (5.39), we take one step ahead

$$z_{k+1}^1 = f^1\left(x_k^1\right) + B^1\left(x_k^1\right) x_k^2 - x_{\delta,k+1}^1. \tag{5.40}$$

Equation (5.40) is viewed as a block with state z_k^1 and the state x_k^2 is considered as

a pseudo-control input, where desired dynamics can be imposed, which can be solved

with the anticipation of the desired dynamics for (5.40) as follows:

$$\begin{aligned}
z_{k+1}^1 &= f^1\left(x_k^1\right) + B^1\left(x_k^1\right) x_k^2 - x_{\delta,k+1}^1 \\
&= f^1\left(z_k^1\right) + B^1\left(z_k^1\right) z_k^2 \tag{5.41}
\end{aligned}$$

Then, x_k^2 is calculated as

$$x_{\delta,k}^2 = \left(B^1\left(x_k^1\right)\right)^{-1}\left(x_{\delta,k+1}^1 - f^1\left(x_k^1\right) + f^1\left(z_k^1\right) + B^1\left(z_k^1\right) z_k^2\right). \tag{5.42}$$

Note that the calculated value of state $x_{\delta,k}^2$ in (5.42) is not the real value of such a

state; instead, it represents the desired behavior for x_k^2. To avoid misunderstandings

the desired value for x_k^2 is referred to as $x_{\delta,k}^2$ in (5.42). Hence, equality (5.41) is

satisfied by substituting the pseudo-control input for x_k^2 in (5.41) as $x_k^2 = x_{\delta,k}^2$, obtaining

$z_{k+1}^1 = f^1\left(z_k^1\right) + B^1\left(z_k^1\right) z_k^2$. The same procedure is used for each subsequent block.

Proceeding the same way as for the first block, a second variable in the new

coordinates is defined as

$$z_k^2 = x_k^2 - x_{\delta,k}^2.$$

Taking one step ahead in z_k^2 yields

$$\begin{aligned}
z_{k+1}^2 &= x_{k+1}^2 - x_{\delta,k+1}^2 \\
&= f^2\left(x_k^1, x_k^2\right) + B^2\left(x_k^1, x_k^2\right) x_k^3 - x_{\delta,k+1}^2.
\end{aligned}$$

The desired dynamics for this block are imposed as

$$
\begin{aligned}
z^2_{k+1} &= f^2\left(x^1_k, x^2_k\right) + B^2\left(x^1_k, x^2_k\right) x^3_k - x^2_{\delta,k+1} \\
&= f^1\left(z^1_k\right) + B^2\left(z^1_k, z^2_k\right) z^2_k.
\end{aligned}
\tag{5.43}
$$

These steps are taken iteratively. At the last step, the known desired variable is $x^r_{\delta,k}$, and the last new variable is defined as

$$
z^r_k = x^r_k - x^r_{\delta,k}.
$$

As usually, taking one step ahead yields

$$
z^r_{k+1} = f^r\left(x_k\right) + B^r\left(x_k\right)\alpha(x_k) - x^r_{\delta,k+1}.
\tag{5.44}
$$

Finally, the desired dynamics for this last block are imposed as

$$
\begin{aligned}
z^r_{k+1} &= f^r\left(x_k\right) + B^r\left(x_k\right)\alpha(x_k) - x^r_{\delta,k+1} \\
&= f^r\left(z_k\right) + B^r\left(z_k\right)\beta(z_k)
\end{aligned}
\tag{5.45}
$$

which is achieved with

$$
\alpha(x_k) = \left(B^r\left(x_k\right)\right)^{-1}\left(x^r_{\delta,k+1} - f^r\left(x_k\right) + f^r\left(z_k\right) + B^r\left(z_k\right)\beta(z_k)\right)
\tag{5.46}
$$

where $\beta(z_k)$ is proposed as

$$
\beta(z_k) = -\frac{1}{2}\left(R + \frac{1}{2}g^T\left(z_k\right)P_k\, g(z_k)\right)^{-1} g^T\left(z_k\right)P_k\, f(z_k).
\tag{5.47}
$$

Then, system (5.38) can be presented in the new variables $z = \left[z^{1T} z^{2T} \cdots z^{rT} \right]$ as

$$z_{k+1}^1 = f^1\left(z_k^1\right) + B^1\left(z_k^1\right) z_k^2$$

$$\vdots$$

$$z_{k+1}^{r-1} = f^{r-1}\left(z_k^1, z_k^2, \ldots, z_k^{r-1}\right) \tag{5.48}$$

$$+ B^{r-1}\left(z_k^1, z_k^2, \ldots, z_k^{r-1}\right) z_k^r$$

$$z_{k+1}^r = f^r\left(z_k\right) + B^r\left(z_k\right) \beta\left(z_k\right).$$

For system (5.48), we establish the following lemma.

Lemma 5.3

Consider that the equilibrium point $x_k = 0$ of system (5.38) is asymptotically stable by means of the control law (5.13), as established in Theorem 5.1. Then the closed-loop solution of transformed system (5.48) with control law (5.47) is (globally) asymptotically stable along the desired trajectory $x_{\delta,k}$, where $P_k = p_k P$, and p_k is calculated using the proposed SG scheme. Moreover, control law (5.47) is inverse optimal in the sense that it minimizes the cost functional

$$\mathscr{J} = \sum_{k=0}^{\infty} \left(l(z_k) + \beta^T(z_k) R \beta(z_k) \right) \tag{5.49}$$

with

$$l(z_k) = -\overline{V}(z_k) \geq 0. \tag{5.50}$$

■

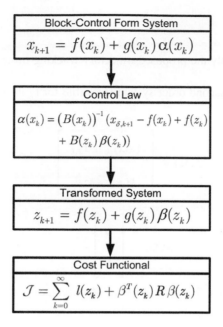

FIGURE 5.11   Transformation scheme and inverse optimal control for the transformed

system.

PROOF

Since the equilibrium point $x_k = 0$ for system (5.38) with a control law of the form

given in (5.13) is globally asymptotically stable, then the equilibrium point $z_k = 0$

(tracking error) for system (5.48) with control law (5.47) is globally asymptotically

stable for the transformed system along the desired trajectory $x_{\delta,k}$.

The minimization of the cost functional is established similarly as in Theorem

5.1, and hence it is omitted. Figure 5.11 displays the proposed transformation and

optimality scheme.   ■

5.3.1 EXAMPLE

In this example, we apply the proposed trajectory tracking inverse optimal control

law for a discrete-time second order system (unstable for $u_k = 0$) of the form (2.1)

with

$$f(x_k) = \begin{bmatrix} 1.5x_{1,k} + x_{2,k} \\ \\ x_{1,k} + 2x_{2,k} \end{bmatrix} \tag{5.51}$$

and

$$g(x_k) = \begin{bmatrix} 0 \\ \\ 1 \end{bmatrix}. \tag{5.52}$$

In accordance with Section 5.3, control law (5.46) becomes

$$\alpha(x_k) = x^1_{\delta,k+2} - 3.25x_{1,k} - 1.5x_{2,k} + 5.25z_{1,k} + 3.5z_{2,k} + 2\beta(z_k) \tag{5.53}$$

where $\beta(z_k)$ is given in (5.47), for which $P = p_k P'$, with

$$P' = \begin{bmatrix} 15 & 7 \\ \\ 7 & 15 \end{bmatrix}$$

which is sufficient to ensure asymptotic stability along the desired trajectory $(x_{\delta,k})$;

$R = 0.01$ and p_k are determined by the SG algorithm with initial condition $p_0 = 0.1$.

Figure 5.12 presents the trajectory tracking for first block $x_{1,k}$, where the solid line

$(x_{\delta,k})$ is the reference signal and the dashed line is the evolution of $x_{1,k}$. The control

signal is also displayed. Figure 5.13 displays the SG algorithm time evolution p_k and

the respective evaluation of the cost functional $\mathcal{J}$.

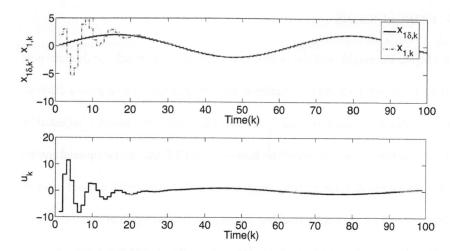

FIGURE 5.12 (**SEE COLOR INSERT**)   Tracking performance of x_k for the system in

block-control form.

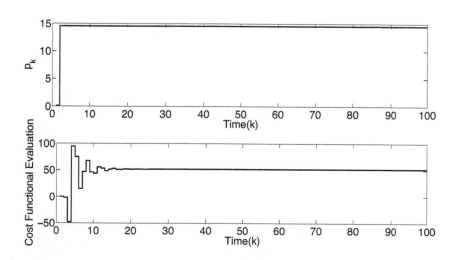

FIGURE 5.13 p_k and $\mathscr{J}$ time evolution for trajectory tracking.

5.4 CONCLUSIONS

This chapter has established the inverse optimal control approach for discrete-time nonlinear systems. To avoid the solution of the HJB equation, we propose a discrete-time CLF in a quadratic form, which depends on a parameter which is adjusted by means of the speed-gradient algorithm. Based on this CLF, the inverse optimal control strategy is synthesized. Then, these results are extended to establish a speed-gradient inverse optimal for trajectory tracking. Finally, an inverse optimal control scheme was presented to achieve trajectory tracking for nonlinear systems in block-control form. Simulation results illustrate the effectiveness of the proposed control schemes.

6 Neural Inverse Optimal Control

This chapter discusses the combination of Section 2.5, Section 3.1, Section 3.2, and Section 5.1. The results of this combination are presented in Section 6.1 to achieve stabilization and trajectory tracking for uncertain nonlinear systems, by using a RHONN scheme to model uncertain nonlinear systems, and then applying the inverse optimal control methodology. The training of the neural network is performed on-line using an extended Kalman filter. Section 6.2 establishes a block transformation for the neural model in order to solve the inverse optimal trajectory tracking as a stabilization problem for block-control form nonlinear systems. Examples illustrate the applicability of the proposed control techniques.

Stabilization and trajectory tracking results can be applied to disturbed nonlinear systems, which can be modeled by means of a neural identifier as presented in Section 2.5, obtaining a robust inverse optimal controller. Two procedures to achieve robust trajectory tracking with the neural model are presented. First, based on the passivity approach, we propose a neural inverse optimal controller which uses a CLF with a global minimum on the desired trajectory. Second, a block transformation for a neural identifier is applied in order to obtain an error system on the desired reference, and then a neural inverse optimal stabilization control law for the error resulting system

is synthesized.

6.1 NEURAL INVERSE OPTIMAL CONTROL SCHEME

First, the stabilization problem for discrete-time nonlinear systems is discussed.

6.1.1 STABILIZATION

As described in Section 2.5, for neural identification of (2.33) a series-parallel neural

model (2.40) can be used. Then, for this neural model, stabilization results established

in Chapter 3 are applied as follows.

Model (2.40) can be represented as a system of the form (2.27)

$$x_{k+1} = f(x_k) + g(x_k) u_k$$

and if there exists $P = P^T > 0$ satisfying condition (3.4), this system can be asymp-

totically stabilized by the inverse optimal control law

$$\alpha(x_k) = -\left(I_m + \frac{1}{2}g^T(x_k)Pg(x_k)\right)^{-1} g^T(x_k)Pf(x_k)$$

in accordance with Theorem 3.1.

An example illustrates the previously mentioned results.

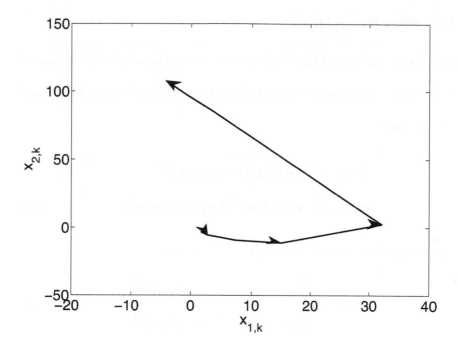

FIGURE 6.1   Unstable system phase portrait.

6.1.2   EXAMPLE

Let us consider a discrete-time second order nonlinear system (unstable for $u_k = 0$)

of the form (2.27) with

$$f(\chi_k) = \begin{bmatrix} 0.5\,\chi_{1,k}\,sin(0.5\,\chi_{1,k}) + 0.2\,\chi_{2,k}^2 \\ 0.1\,\chi_{1,k}^2 + 1.8\,\chi_{2,k} \end{bmatrix} \tag{6.1}$$

and

$$g(\chi_k) = \begin{bmatrix} 0 \\ 2 + 0.1\,cos(\chi_{2,k}) \end{bmatrix}. \tag{6.2}$$

The phase portrait for this unstable open-loop ($u_k = 0$) system with initial condi-

tions $\chi_0 = [2 \ \ -2]^T$ is displayed in Figure 6.1.

6.1.2.1 Neural Network Identifier

Let us assume that system (2.27) with (6.1)–(6.2) to be unknown. In order to identify

this uncertain system, from (2.37) and (2.40), we propose the following series–parallel

neural network:

$$x_{1,k+1} \;=\; w_{11,k} S(\chi_1) + w_{12,k} (S(\chi_2))^2$$

$$x_{1,k+1} \;=\; w_{21,k} (S(\chi_1))^2 + w_{22,k} S(\chi_2) + w_2' u_k \qquad (6.3)$$

which can be rewritten as $x_{k+1} = f(x_k) + g(x_k) u_k$, where

$$f(x_k) = \begin{bmatrix} w_{11,k} S(\chi_1) + w_{12,k} (S(\chi_2))^2 \\ w_{21,k} (S(\chi_1))^2 + w_{22,k} S(\chi_2) \end{bmatrix} \qquad (6.4)$$

and

$$g(x_k) = \begin{bmatrix} 0 \\ w_2' \end{bmatrix} \qquad (6.5)$$

with $w_2' = 0.8$. The initial conditions for the adjustable weights are selected as Gaus-

sian random values, with zero mean and a standard deviation of 0.333; $\eta_1 = \eta_2 = 0.99$,

$P_1 = P_2 = 10 I_2$, where I_2 is the 2×2 identity matrix; $Q_1 = Q_2 = 1300 I_2$, $R_1 = 1000$,

and $R_2 = 4500$.

6.1.2.2 Control Synthesis

According to (3.1), the inverse optimal control law is formulated as

$$\alpha(x_k) = -\left(1 + \frac{1}{2} g^T(x_k) P g(x_k)\right)^{-1} g^T(x_k) P f(x_k) \qquad (6.6)$$

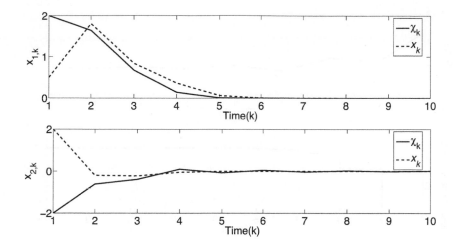

FIGURE 6.2   Stabilized system time response.

with P as

$$P = \begin{bmatrix} 0.0005 & 0.0319 \\ 0.0319 & 3.2942 \end{bmatrix}.$$

Figure 6.2 presents the stabilization time response for x_k with initial conditions

$\chi_0 = [2 \ -2]^T$. Initial conditions for RHONN are $x_0 = [0.5 \ 2]^T$. Figure 6.3 displays

the applied inverse optimal control law (6.6), which achieves asymptotic stability;

this figure also includes the cost functional values.

6.1.3   TRAJECTORY TRACKING

The tracking of a desired trajectory, defined in terms of the plant state χ_i formulated

as (2.33), can be established as the following inequality:

$$\|\chi_{i\delta} - \chi_i\| \leq \|\chi_i - x_i\| + \|\chi_{i\delta} - x_i\| \tag{6.7}$$

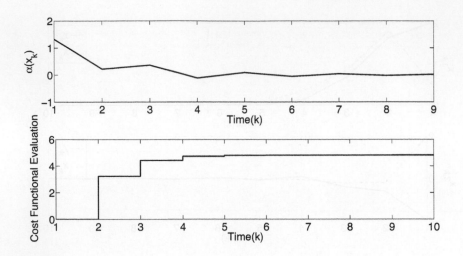

FIGURE 6.3   Control law and cost functional values.

where $\|\cdot\|$ stands for the Euclidean norm, $\chi_{i\delta}$ is the desired trajectory signal, which is assumed smooth and bounded. Inequality (6.7) is valid considering the separation principle for discrete-time nonlinear systems [68], and based on (6.7), the tracking of a desired trajectory can be divided into the following two requirements.

REQUIREMENT 6.1

$$\lim_{k \to \infty} \|\chi_i - x_i\| \leq \zeta_i \tag{6.8}$$

with ζ_i a small positive constant.

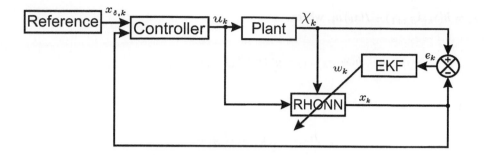

FIGURE 6.4   Control scheme.

REQUIREMENT 6.2

$$\lim_{k \to \infty} \|\chi_{i\delta} - x_i\| = 0. \tag{6.9}$$

In order to fulfill Requirement 6.1, an on-line neural identifier based on (2.37) is proposed to ensure (6.8) [116], while Requirement 6.2 is guaranteed by a discrete-time controller developed using the inverse optimal control technique. A general control scheme is shown in Figure 6.4.

Trajectory tracking is illustrated as established in Section 3.2 for the neural scheme as follows.

6.1.3.1   Example

In accordance with Theorem 3.2, in order to achieve trajectory tracking for $x_{k+1} = f(x_k) + g(x_k) u_k$ with $f(x_k)$ and $g(x_k)$ as defined by the neural model in (6.4) and (6.5), respectively, then the control law is established in (3.28) as $u_k = -y_k$, for which

$y_k = h(x_k, x_{\delta,k+1}) + J(x_k) u_k$, where

$$h(x_k, x_{\delta,k+1}) = g^T(x_k) \overline{P}(f(x_k) - x_{\delta,k+1})$$

and

$$J(x_k) = \frac{1}{2} g^T(x_k) \overline{P} g(x_k)$$

with the signal reference for $x_{2,k}$ as

$$x_{2\delta,k} = 2 \sin(0.075 k) \text{ rad}$$

and reference $x_{1\delta,k}$ defined accordingly with the system dynamics. Hence, we adjust

gain matrix $\overline{P} = K^T P K$ for (3.28) in order to achieve trajectory tracking for $x_k =$

$[x_{1,k}\, x_{2,k}]^T$.

Figure 6.5 presents trajectory tracking for x_k with

$$P = \begin{bmatrix} 0.0484 & 0.0387 \\ 0.0387 & 0.0484 \end{bmatrix} ; \; K = \begin{bmatrix} 0.100 & 0.00 \\ 0.00 & 8.25 \end{bmatrix}.$$

Figure 6.6 presents the applied control signal to achieve trajectory tracking; it also

displays the cost functional values, which increase because the control law is different

from zero in order to achieve trajectory tracking.

6.1.4 APPLICATION TO A SYNCHRONOUS GENERATOR

In this section, we apply the speed-gradient inverse optimal control technique to a

discrete-time neural model identifying a synchronous generator [4]. Figure 6.7 depicts

a power synchronous generator. The goal of power system stabilization is to produce

stable, reliable, and robust electrical energy production and distribution. One task to

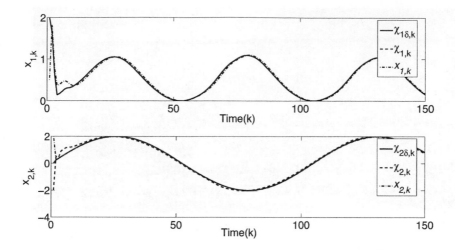

FIGURE 6.5   System time response for trajectory tracking.

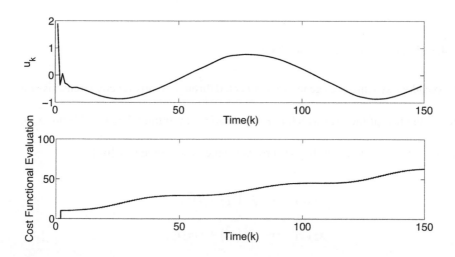

FIGURE 6.6   Control law for trajectory tracking and cost functional values.

FIGURE 6.7 (**SEE COLOR INSERT**)   Synchronous generator.

achieve this goal is to reduce the adverse effects of mechanical oscillations, which

can induce premature degradation of mechanical and electrical components as well

as dangerous operation of the system (e.g., blackouts) [9].

6.1.4.1   Synchronous Generator Model

We consider a synchronous generator connected through purely reactive transmission

lines to the rest of the grid which is represented by an infinite bus (see Figure 6.8).

The discrete-time model of the synchronous generator is given by [64]

$$x_{1,k+1} = f^1\left(\bar{x}_k^1\right) + \tau x_{2,k}$$

$$x_{2,k+1} = f^2\left(\bar{x}_k^2\right) + \tau m_2 x_{3,k}$$

$$x_{3,k+1} = f^3\left(\bar{x}_k^3\right) + \tau m_6 u_k \tag{6.10}$$

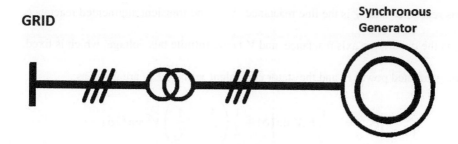

GRID

Synchronous Generator

FIGURE 6.8   Synchronous generator connected to infinite bus.

with

$$f^1\left(\bar{x}_k^1\right) = x_{1,k}$$

$$f^2\left(\bar{x}_k^2\right) = x_{2,k} + \tau\left[m_1 + \left(m_2 E_q^{'*} + m_3\cos\left(\tilde{x}_1\right)\right)\sin\left(\tilde{x}_1\right)\right]$$

$$f^3\left(\bar{x}_k^3\right) = x_{3,k} + \tau\left[m_4\left(x_{3,k} + E_q^{'*}\right) + m_5\cos\left(\tilde{x}_1\right) + m_6 E_{fd}^*\right]$$

and $\tilde{x}_1 = x_{1,k} + \delta^*$, $m_1 = \frac{T_m}{M}$, $m_2 = \frac{-V}{MX_d'}$, $m_3 = \frac{V^2}{M}\left(\frac{1}{X_d'} - \frac{1}{X_q}\right)$, $m_4 = -\frac{X_d}{T_{do}'X_d'}$, $m_5 = -\left(\frac{X_d'-X_d}{T_{do}'X_d'}\right)V$, $m_6 = \frac{1}{T_{do}'}$, with

$$x_1 \quad := \quad \Delta\delta = \delta - \delta^*$$

$$x_2 \quad := \quad \Delta\omega = \omega - \omega^* \qquad\qquad (6.11)$$

$$x_3 \quad := \quad \Delta E_q' = E_q' - E_q^{'*}$$

where δ is the generator rotor angle referred to the infinite bus (also called power angle), ω is the rotor angular speed, and E_q' is the stator voltage which is proportional to flux linkages; τ is the sampling time, M is the per unit inertia constant, T_m is the constant mechanical power supplied by the turbine, and T_{do}' is the transient open circuit time constant. $X_d = x_d + x_L$ is the augmented reactance, where x_d is the direct axis reactance and x_L is the line reactance, X'_d is the transient augmented reactance, X_q is the quadrature axis reactance, and V is the infinite bus voltage, which is fixed. The generated power P_g and the stator equivalent voltage E_{fd} are given as

$$
\begin{aligned}
P_g &= \frac{1}{X'_d} E'_q V \sin(\delta) + \frac{1}{2} \left(\frac{1}{X_q} - \frac{1}{X'_d} \right) V^2 \sin(2\delta) \\
E_{fd} &= \frac{\omega_s M_f}{\sqrt{2} R_f} v_f
\end{aligned}
$$

respectively, where v_f is the scaled field excitation voltage, M_f is the mutual inductance between stator coils, R_f is the field resistance, and ω_s is the synchronous speed. As in [64], we only consider the case where the dynamics of the damper windings are neglected.

Through this work, the analysis and design are done around an operation point $\left(\delta^*, \omega^*, E'^*_q \right)$, which is obtained for a stator field equivalent voltage $E^*_{fd} = 1.1773$ as proposed in [64], for which $\delta^* = 0.870204$, $\omega^* = 1$, and $E'^*_q = 0.822213$. The sampling time is selected as $\tau = 0.01$.

The parameters of the plant model used for simulation are given in the Table 6.1.

6.1.4.2 Neural Identification for the Synchronous Generator

In order to identify the discrete-time synchronous generator model (6.10), we propose a RHONN as follows:

$$
\begin{aligned}
\widehat{x}_{1,k+1} &= w_{11,k} S\left(x_{1,k}\right) + w_{12,k} S\left(x_{2,k}\right) \\
\widehat{x}_{2,k+1} &= w_{21,k} S\left(x_{1,k}\right)^6 + w_{22,k} S\left(x_{2,k}\right)^2 + w_{23,k} S\left(x_{3,k}\right) \qquad (6.12) \\
\widehat{x}_{3,k+1} &= w_{31,k} S\left(x_{1,k}\right)^2 + w_{32,k} S\left(x_{2,k}\right) + w_{33,k} S\left(x_{3,k}\right)^2 + w_{34} u_k
\end{aligned}
$$

TABLE 6.1

Model parameters (per unit).

PARAMETER	VALUE		PARAMETER	VALUE
T_m	1		X_q	0.9
M	0.033		X_d	0.9
ω_s	0.25		X_d'	0.3
T_{do}'	0.033		V	1.0

where $\hat{x}_i$ estimates x_l ($i = 1, 2, 3$), and w_{34} is a fixed parameter in order to ensure the controllability of the neural identifier [94], which is selected as $w_{34} = 0.5$. System (6.13) can be rewritten as

$$x_{k+1} = f(x_k) + g(x_k)u_k$$

with

$$f(x_k) = \begin{bmatrix} w_{11,k}S(x_{1,k}) + w_{12,k}S(x_{2,k}) \\ w_{21,k}S(x_{1,k})^6 + w_{22,k}S(x_{2,k})^2 + w_{23,k}S(x_{3,k}) \\ w_{31,k}S(x_{1,k})^2 + w_{32,k}S(x_{2,k}) + w_{33,k}S(x_{3,k})^2 \end{bmatrix} \quad (6.13)$$

and

$$g(x_k) = \begin{bmatrix} 0 \\ 0 \\ w_{34} \end{bmatrix}. \quad (6.14)$$

The training is performed on-line, using a series-parallel configuration, with the EKF

learning algorithm (2.41). All the NN states are initialized in a random way as well as

the weights vectors. It is important to remark that the initial conditions of the plant are

completely different from the initial conditions of the NN. The covariance matrices

are initialized as diagonals, and the nonzero elements are $P_1(0) = P_2(0) = 10,000$;

$Q_1(0) = Q_2(0) = 5000$ and $R_1(0) = R_2(0) = 10,000$, respectively.

6.1.4.3 Control Synthesis

For system (2.1), with $f(x_k)$ and $g(x_k)$ as defined in (6.13) and (6.14), respectively,

the proposed inverse optimal control law (5.2) is formulated as

$$u_k^* = -\frac{1}{2}\left(R + \frac{1}{2}g^T(x_k)P_k g(x_k)\right)^{-1} g^T(x_k)P_k f(x_k)$$

where R is a constant matrix selected as $R = 1$, and P_k is defined as $P_k = p_k P'$, for

which p_k is calculated by the SG algorithm, and P' is a symmetric and positive definite

matrix selected as

$$P' = \begin{bmatrix} 1.5 & 1.0 & 0.5 \\ 1.0 & 1.5 & 1.0 \\ 0.5 & 1.0 & 1.5 \end{bmatrix}. \tag{6.15}$$

6.1.4.4 Simulation Results

Initially, results for the inverse optimal controller based on the plant model are por-

trayed. Figure 6.9 displays the solutions of $\left[\delta, \omega, E_q'\right]^T$, which are based in the system

(6.10) and (6.11), with initial conditions $[0.77, 0.10, 0.85]^T$, respectively. Addition-

ally, in order to illustrate the robustness of the proposed controller, the simulation

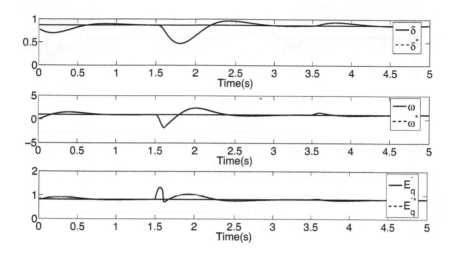

FIGURE 6.9 Time evolution for x_k.

stages are indicated as follows:

Stage 1: Nominal parameters, as the given in Table 6.1, are used at the beginning of the simulation.

Stage 2: A short circuit fault occurs at 1.5 seconds, which is equivalent to changing the augmented reactance X_d from $X_d = 0.9$ to $X_d = 0.1$.

Stage 3: The short circuit fault is removed at 1.6 seconds.

Stage 4: A disturbance in the mechanical power is carried out by changing T_m from $T_m = 1$ to $T_m = 1.2$ at 3.5 seconds.

Stage 5: The disturbance is removed at 3.6 seconds.

Stage 6: The system is in a post-fault and post-disturbance state.

Figure 6.10 includes the applied inverse optimal control law, the time evolution for parameter p_k, and the cost functional evaluation.

Finally, the application of the proposed inverse optimal neural controller based on

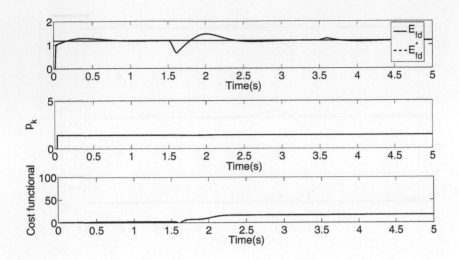

FIGURE 6.10   Time evolution for control law u_k, parameter p_k, and the cost functional.

the neural identifier is presented. Figure 6.11 displays the solution of $\left[\delta, \omega, E_q'\right]^T$,

for the neural identifier (6.13), with the initial conditions given as $[0.77, 0.10, 0.85]^T$,

respectively. In order to illustrate the robustness of the proposed neural controller, the

same simulation stages, as explained above, were carried out for this neural scheme.

Figure 6.12 includes the applied inverse optimal neural control law, the time evolution

for parameter p_k, and the cost functional evaluation. The control goal is to guaran-

tee that all the state variables values stay at their equilibrium point (regulation), as

illustrated by the simulations.

6.1.5   COMPARISON

In order to compare the proposed control schemes, Table 6.2 is included, and is

described as follows. The controllers used in this comparison are: 1) speed-gradient

inverse optimal control (SG-IOC) and 2) speed-gradient inverse optimal neural control

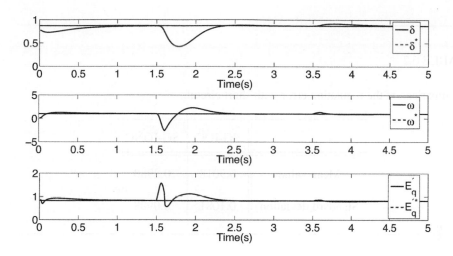

FIGURE 6.11   Time evolution for x_k using the neural identifier.

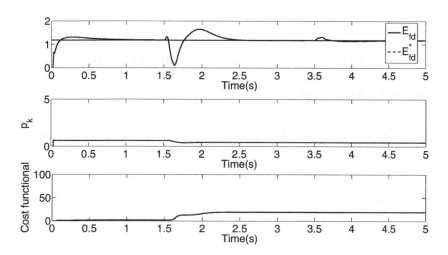

FIGURE 6.12   Time evolution for control law u_k, parameter p_k, and the cost functional using

the neural identifier.

TABLE 6.2

Comparison of the regulation error with short circuit.

	SG-IOC	SG-IONC
Mean value	0.9018	0.8980
Standard deviation	0.1081	0.0988

(SG-IONC).

It is important to note that for the SG-IOC, it is required to know the exact plant

model (structure and parameters) for the controller synthesis, which in many cases is

a disadvantage [95].

6.2   BLOCK-CONTROL FORM: A NONLINEAR SYSTEMS

   PARTICULAR CLASS

In this section, we develop a neural inverse optimal control scheme for systems which

have a special state representation referred as the block-control (BC) form [73].

6.2.1 BLOCK TRANSFORMATION

Let us consider that system (2.33), under an appropriate nonsingular transformation, can be rewritten as the following BC form with r blocks:

$$\chi_{i_1,k+1} = f_{i_1}(\chi_{i_1,k}) + B_{i_1}(\chi_{i_1,k})\chi_{i_1,k} + \Gamma_{i_1,k}$$

$$\chi_{i_2,k+1} = f_{i_2}(\chi_{i_1,k}, \chi_{i_2,k}) + B_{i_2}(\chi_{i_1,k}, \chi_{i_2,k})\chi_{i_3,k} + \Gamma_{i_2,k}$$

$$\vdots \qquad\qquad (6.16)$$

$$\chi_{i_r,k+1} = f_{i_r}(\chi_k) + B_{i_r}(\chi_k)u_k + \Gamma_{i_r,k}$$

where $\chi_k \in \mathbb{R}^n$ is the system state, $\chi_k = \begin{bmatrix} \chi_{i_1,k}^T & \chi_{i_2,k}^T & \cdots & \chi_{i_2,k}^T \end{bmatrix}^T$; $i = 1,\ldots,n_r$; $u_k \in \mathbb{R}^{m_r}$ is the input vector. We assume that f_{i_j}, B_{i_j}, and Γ_{i_j} are smooth functions, $j = 1,\ldots,r$, $f_{i_j}(0) = 0$, and $rank\{B_{i_j}(\chi_k)\} = m_j \ \forall \chi_k \neq 0$. The unmatched and matched disturbance terms are represented by Γ_i. The whole system order is $n = \sum_{j=1}^{r} n_j$.

To identify (6.16), we propose a neural network with the same BC structure, consisting of r blocks as follows:

$$x_{i_1,k+1} = W_{i_1,k}\rho_{i_1}(\chi_{i_1,k}) + W_{i_1}'\chi_{i_2,k}$$

$$x_{i_2,k+1} = W_{i_2,k}\rho_{i_2}(\chi_{i_1,k}, \chi_{i_2,k}) + W_{i_2}'\chi_{i_3,k} \qquad (6.17)$$

$$\vdots$$

$$x_{i_r,k+1} = W_{i_r,k}\rho_{i_r}(\chi_{i_1,k},\ldots,\chi_{i_r,k}) + W_{i_r}'u_k$$

where $x_k = \begin{bmatrix} x_{i_1}^T x_{i_2}^T \cdots x_{i_r}^T \end{bmatrix}^T \in \mathbb{R}^n$, $x_{i_r} \in \mathbb{R}^{n_r}$ denotes the i-th neuron system state corresponding to the r-th block; $i = 1,\ldots,n_r$; $W_{i_1,k} = [w_{1_1,k}^T \ w_{2_1,k}^T \ \cdots \ w_{n_{r1},k}^T]^T$ is the on-line adjustable weight matrix, and $W_{i_r}' = [w_{1_1}'^T \ w_{2_1}'^T \ \cdots \ w_{n_{r1}}'^T]^T$ is the fixed weight matrix; n_r

denotes the order for each r-th block, and the whole system order becomes $n = \sum_{j=1}^{r} n_j$.

First, we define the tracking error as

$$z_{i_1,k} = x_{i_1,k} - \chi_{i_1\delta,k} \tag{6.18}$$

where $\chi_{i_1\delta,k}$ is the desired trajectory signal.

Once the first new variable (6.18) is defined, one step ahead is taken as

$$z_{i_1,k+1} = W_{i_1,k}\rho_{i_1}(\chi_{i_1,k}) + W'_{i_1}\chi_{i_2,k} - \chi_{i_1\delta,k+1}. \tag{6.19}$$

Equation (6.19) is viewed as a block with state $z_{i_1,k}$ and the state $\chi_{i_2,k}$ is considered as a pseudo-control input, where desired dynamics can be imposed, which can be solved with the anticipation of the desired dynamics for this block as follows:

$$z_{i_1,k+1} = W_{i_1,k}\rho_{i_1}(\chi_{i_1,k}) + W'_{i_1}\chi_{i_2,k} - \chi_{i_1\delta,k+1} = K_{i_1}z_{i_1,k} \tag{6.20}$$

where $K_{i_1} = diag\{k_{11},\ldots,k_{n_11}\}$ with $|k_{q1}| < 1$, $q = 1,\ldots,n_1$, in order to assure stability for block (6.20). From (6.20), $\chi_{i_2,k}$ is calculated as

$$\chi_{i_2\delta,k} = \left(W'_{i_1}\right)^{-1}\left(-W_{i_1,k}\rho_{i_1}(\chi_{i_1,k}) + \chi_{i_1\delta,k+1}\right.$$

$$\left. + K_{i_1}z_{i_1,k}\right). \tag{6.21}$$

Note that the calculated value for state $\chi_{i_2\delta,k}$ in (6.21) is not the real value for such a state; instead, it represents the desired behavior for $\chi_{i_2,k}$. Hence, to avoid confusion this desired value of $\chi_{i_2,k}$ is referred to as $\chi_{i_2\delta,k}$ in (6.21).

Proceeding in the same way as for the first block, a second variable in the new coordinates is defined as

$$z_{i_2,k} = x_{i_2,k} - \chi_{i_2\delta,k}.$$

Taking one step ahead in $z_{i_2,k}$ yields

$$
\begin{aligned}
z_{i_2,k+1} &= x_{i_2,k+1} - x_{i_2\delta,k+1} \\
&= W_{i_2,k}P_{i_2}(\chi_{i_1,k},\chi_{i_2,k}) + W'_{i_2}\chi_{i_3,k} - \chi_{i_2\delta,k+1}.
\end{aligned}
$$

The desired dynamics for this block are imposed as

$$
\begin{aligned}
z_{i_2,k+1} &= W_{i_2,k}P_{i_2}(\chi_{i_1,k},\chi_{i_2,k}) + W'_{i_2}\chi_{i_3,k} - \chi_{i_2\delta,k+1} \\
&= K_{i_2}z_{i_2,k} \qquad\qquad\qquad\qquad\qquad\qquad (6.22)
\end{aligned}
$$

where $K_{i_2} = diag\{k_{12},\ldots,k_{n_2 2}\}$ with $|k_{q2}| < 1$, $q = 1,\ldots,n_2$.

These steps are taken iteratively. At the last step, the known desired variable is $\chi^r_{r\delta,k}$, and the last new variable is defined as

$$
z_{i_r,k} = x_{i_r,k} - \chi_{i_r\delta,k}.
$$

As usual, taking one step ahead yields

$$
z_{i_r,k+1} = W_{i_r,k}P_{i_r}(\chi_{i_1,k},\ldots,\chi_{i_r,k}) + W'_{i_r}u_k - \chi_{i_r\delta,k+1}. \qquad (6.23)
$$

System (6.17) can be represented in the new variables as

$$
z_{i_1,k+1} = K_1 z_{i_1,k} + W'_{i_1}z_{i_2,k}
$$

$$
z_{i_2,k+1} = K_{i_2}z_{i_2,k} + W'_{i_2}z_{i_3,k}
$$

$$
\vdots \qquad\qquad\qquad\qquad\qquad\qquad (6.24)
$$

$$
z_{i_r,k+1} = W_{i_r,k}P_{i_r}(\chi_{i_1,k},\ldots,\chi_{i_r,k}) - \chi_{i_r\delta,k+1} + W'_{i_r}u_k.
$$

6.2.2 BLOCK INVERSE OPTIMAL CONTROL

In order to achieve trajectory tracking along $\chi_{\delta,k}$, the transformed system (6.24) is

stabilized at its origin. System (6.24) can be presented in a general form as

$$z_{k+1} = f(z_k) + g(z_k) u_k \qquad (6.25)$$

where $z_k = \left[z_{i_1,k}^T z_{i_2,k}^T \cdots z_{i_r,k}^T \right]^T$.

Then, in order to achieve stabilization for (6.25) along the desired trajectory, we

apply the inverse optimal control law (3.1) as

$$u_k = \alpha(z_k) = - \left[I_m + J(z_k) \right]^{-1} h(z_k)$$

with

$$h(z_k) = g(z_k)^T P f(z_k) \qquad (6.26)$$

and

$$J(z_k) = \frac{1}{2} g(z_k)^T P g(z_k). \qquad (6.27)$$

6.2.3 APPLICATION TO A PLANAR ROBOT

In order to illustrate the application of the control scheme developed in the previous

section, we present the position trajectory tracking for a planar robot by using the

block-control system representation as established in Section 6.2.

6.2.3.1 Robot Model Description

After a Euler discretization of the robot dynamics, the discrete-time robot model is

described by (3.39), which has the BC form.

REMARK 6.1 The structure of system (3.39) is used to design the neural network identifier. The parameters of system (3.39) are assumed to be unknown for control synthesis. ∎

REMARK 6.2 For system (3.39), $n = 4, r = 2, n_1 = 2, n_2 = 2$. To identify this system, the series-parallel model (2.40) is used. ∎

6.2.3.2 Neural Network Identifier

To identify the uncertainty robot model, from (2.37), (2.40), and (6.17), we propose the following series-parallel neural network according to Remark 6.1 and Remark 6.2 as

$$
\begin{bmatrix} x_{1_1,k+1} \\ x_{2_1,k+1} \end{bmatrix} = \begin{bmatrix} w_{1_11,k} S(\chi_{1_1,k}) + w'_{1_1} \chi_{1_2,k} \\ w_{2_11,k} S(\chi_{2_1,k}) + w'_{2_1} \chi_{2_2,k} \end{bmatrix}
$$

$$
\begin{bmatrix} x_{1_2,k+1} \\ x_{2_2,k+1} \end{bmatrix} = \begin{bmatrix} w_{1_21,k} S(\chi_{1_1,k}) + w_{1_22,k} S(\chi_{2_1,k}) \cdots \\ w_{2_21,k} S(\chi_{1_1,k}) + w_{2_22,k} S(\chi_{2_1,k}) \cdots \end{bmatrix}
$$

$$
\begin{aligned}
&+ w_{1_23,k} S(\chi_{1_2,k}) + w_{1_24,k} S(\chi_{2_2,k}) \cdots \\
&+ w_{2_23,k} S(\chi_{1_2,k}) + w_{2_24,k} S(\chi_{2_2,k}) \cdots \\
&+ w'_{1_21} u_{1,k} + w'_{1_22} u_{2,k} \\
&+ w'_{2_21} u_{1,k} + w'_{2_22} u_{2,k}
\end{aligned}
\Bigg]
\tag{6.28}
$$

where $x_{j_1,k}$ identifies to $\chi_{j_1,k}$ and $x_{j_2,k}$ identifies to $\chi_{j_2,k}$; $j = 1,2$; w_{j_rp} are the adjustable weights, p is the corresponding number of adjustable weights; w'_{j_rp} are fixed parameters.

To update the weights, the adaptation algorithm (2.41) is implemented.

6.2.3.3 Control Synthesis

Let us rewrite system (6.28) in an r-block-control form as

$$x_{1,k+1}: \quad = \quad \begin{bmatrix} x_{1_1,k+1} \\ \\ x_{2_1,k+1} \end{bmatrix} = W_{1,k}\rho_1(\chi_{1,k}) + W_1'\chi_{2,k} \tag{6.29}$$

$$x_{2,k+1}: \quad = \quad \begin{bmatrix} x_{1_2,k+1} \\ \\ x_{2_2,k+1} \end{bmatrix} = W_{2,k}\rho_2(\chi_{1,k},\chi_{2,k}) + W_2'u_k$$

with $\chi_{1,k}, \chi_{2,k}, \rho_1, \rho_2, W_{1,k}, W_{2,k}, W_1'$, and W_2' of appropriate dimensions according to

(6.28).

The goal is to force the angle position $x_{1,k}$ to track a desired reference signal $\chi_{1\delta,k}$.

This is achieved by designing a control law as described in Section 6.2.1.

First the tracking error is defined as

$$z_{1,k} = x_{1,k} - \chi_{1\delta,k}.$$

Then using (6.28) and introducing the desired dynamics for $z_{1,k}$ results in

$$z_{1,k+1} \quad = \quad W_{1,k}\rho_1(\chi_{1,k}) + W_1'\chi_{2,k} - \chi_{1\delta,k+1}$$

$$= \quad K_1 z_{1,k} \tag{6.30}$$

where $K_1 = diag\{k_{1_1}, k_{2_1}\}$ with $|k_{1_1}|, |k_{2_1}| < 1$.

The desired value $\chi_{2\delta,k}$ for the pseudo-control input $\chi_{2,k}$ is calculated from (6.30) as

$$\chi_{2\delta,k} = \left(W_1'\right)^{-1}\left(-W_{1,k}\rho_1(\chi_{1,k})\right.$$

$$\left.+\chi_{1\delta,k+1}+K_1 z_{1,k}\right). \tag{6.31}$$

At the second step, we introduce a new variable as

$$z_{2,k} = x_{2,k} - \chi_{2\delta,k}.$$

Taking one step ahead, we have

$$z_{2,k+1} = W_{2,k}\rho_2(\chi_{1,k},\chi_{2,k}) - \chi_{2\delta,k+1}+W_2' u_k. \tag{6.32}$$

Now, system (6.29) in the new variables $z_{1,k}$ and $z_{2,k}$ is represented as

$$z_{1,k+1} = K_1 z_{1,k} + W_1' z_{2,k}$$

$$z_{2,k+1} = W_{2,k}\rho_2(\chi_{1,k},\chi_{2,k}) - \chi_{2\delta,k+1}+W_2' u_k. \tag{6.33}$$

If we rewrite system (6.33) as (6.24)–(6.25), from Section 6.2.2, the proposed control law is given by

$$u_k = \alpha(z_k) = -(1+J(z_k))^{-1} h(z_k) \tag{6.34}$$

where $h(z_k)$ and $J(z_k)$ are defined as in (6.26) and (6.27), respectively.

6.2.3.4 Simulation Results

The initial conditions are given in Table 6.3; the identifier (6.28) and controller parameters are shown in Table 6.4, where I_4 is the 4×4 identity matrix. The sample time is $T = 0.001$ s and the initial conditions for the adjustable weights are selected as Gaussian random values, with zero mean and a standard deviation of 0.333.

TABLE 6.3

Initial conditions of the planar robot and the neural identifier.

PARAMETER	VALUE	PARAMETER	VALUE
$\chi_{1_1,0}$	0.5 rad	$\chi_{2_1,0}$	-0.5 rad
$\chi_{1_2,0}$	0 rad/s	$\chi_{2_2,0}$	0 rad/s
$P_{1_1,0}$	10	$P_{2_1,0}$	10
$P_{1_2,0}$	$10I_4$	$P_{2_1,0}$	$10I_4$
$x_{1_1,0}$	0 rad	$x_{2_1,0}$	0 rad
$x_{1_2,0}$	0 rad/s	$x_{2_2,0}$	0 rad/s

TABLE 6.4

Identifier and controller parameters.

PARAMETER	VALUE		PARAMETER	VALUE
Q_{1_1}	1000		Q_{2_1}	1000
Q_{1_2}	$1000\,I_4$		Q_{2_1}	$1000\,I_4$
R_{1_1}	10,000		R_{2_1}	10,000
R_{1_2}	10,000		R_{2_1}	10,000
w'_{1_1}	0.001		w'_{2_1}	0.001
w'_{1_21}	0.900		w'_{1_22}	0.010
w'_{2_21}	0.010		w'_{2_22}	0.700
k_{1_1}	0.960		k_{2_1}	0.970
η_{1_1}	0.900		η_{1_2}	0.900
η_{2_1}	0.900		η_{2_2}	0.900

The value of matrix P is

$$
P = \begin{bmatrix}
1.1520 & 0.1512 & 0.2880 & 0.0151 \\
0.1512 & 0.0794 & 0.1512 & 0.0198 \\
0.2880 & 0.1512 & 1.1520 & 0.1512 \\
0.0151 & 0.0198 & 0.1512 & 0.0794
\end{bmatrix}.
$$

The reference signals are

$$
\chi_{1_1\delta,k} = 2.0\,sin(1.0kT)\,\text{rad}
$$

$$
\chi_{2_1\delta,k} = 1.5\,sin(1.2kT)\,\text{rad}
$$

which are selected to illustrate the ability of the proposed algorithm to track nonlinear trajectories.

Tracking performances of link 1 and link 2 positions are shown in Figure 6.13. The control signals $u_{1,k}$ and $u_{2,k}$ time responses are displayed in Figure 6.14.

6.3 CONCLUSIONS

This chapter has presented a discrete-time neural inverse optimal control for uncertain nonlinear systems, which is inverse in the sense that it, a posteriori, minimizes a cost functional. We use discrete-time recurrent neural networks to model uncertain nonlinear systems; thus, an explicit knowledge of the plant is not necessary. The proposed approach is successfully applied to implement a controller based on RHONN and the inverse optimal control approach. Simulation results illustrate that the required goal is achieved, i.e., the proposed controller ensures stability or trajectory tracking of the unknown system.

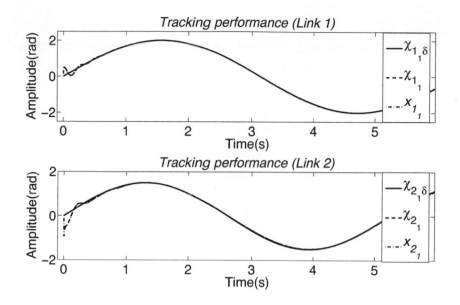

FIGURE 6.13   Planar robot tracking performance.

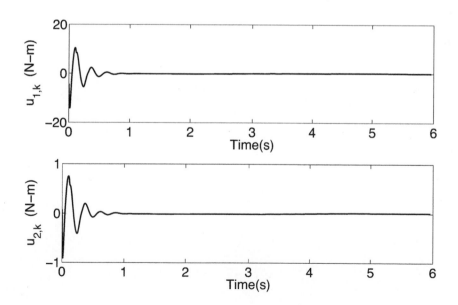

FIGURE 6.14   Control signal time responses.

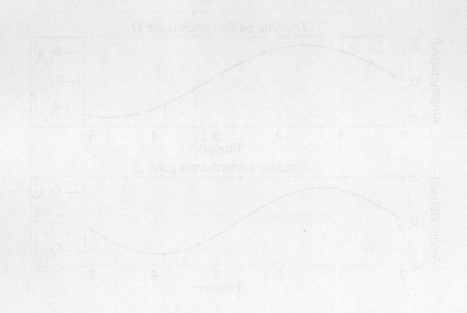

7 Glycemic Control of Type 1 Diabetes Mellitus Patients

The neural inverse optimal control for trajectory tracking is applied to glycemic control of T1DM patients. The proposed control law calculates adequate insulin delivery rates in order to prevent hyperglycemia and hypoglycemia levels in T1DM patients. In Section 7.1 a review of diabetes mellitus is presented. The neural inverse optimal control scheme based on the passivity approach is presented in Section 7.2, while the neural inverse optimal control scheme based on the CLF approach is stated in Section 7.3.

7.1 INTRODUCTION

Type 1 diabetes mellitus is a metabolic disease caused by destruction of insulin producing beta cells in the pancreas. Diabetes mellitus is one of the costliest health problems in the world and one of the major causes of death worldwide. 171 million of people were affected by diabetes mellitus (DM) in 2000 and it is estimated that the number will increase to 366 million in 2030 [133]. According to the International

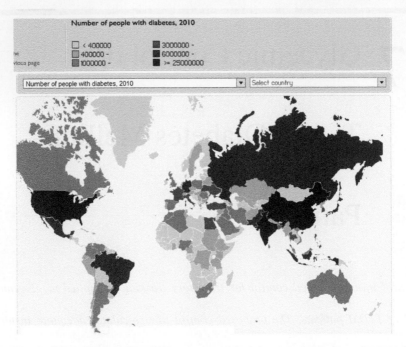

FIGURE 7.1 **(SEE COLOR INSERT)**   People with diabetes in 2009. (From IDF Diabetes

Atlas 4th ed. © International Diabetes Federation, 2009.)

Diabetes Federation, in 2010 the estimated global expenditures to treat and prevent

diabetes and its complications are at least US $376 billion, and this amount will in-

crease to US $490 billion by 2030. Figure 7.1 displays an estimate of people with this

disease. Considering all the problems related to diabetes mellitus, the development

of an artificial pancreas is a challenging task.

Normal blood glucose level is between 90 and 110 mg/dl; T1DM patients can

present with two conditions: hyperglycemia, which means high glucose levels in the

blood plasma (larger than 180 mg/dl) or hypoglycemia (less than 70 mg/dl), which

means a lower than normal level of blood glucose. In order to prevent hyperglycemia

conditions, people with T1DM need injections of insulin. Chronic elevation of blood glucose will eventually lead to tissue damage, which means chronic complications like cardiovascular disease, nephropathy, neuropathy, amputation, and retinopathy [16]. Hypoglycemia has serious short-term complications, such as diabetic coma.

Studies have pointed out that continuous insulin infusion in T1DM patients can reduce or delay the hyperglycemia complications and improve their quality of life [127, 60]. The common insulin therapy consists in providing insulin to the patient intravenously or subcutaneously; it is done usually before meals to prevent rises in glucose levels. To the best of our knowledge, there is not any commercial device capable controlling blood glucose levels in a closed-loop, i.e., a device which measures real-time blood glucose levels, calculates the insulin dose, and provides the insulin to the patient.

In this chapter, attention is drawn to identification and control of the blood glucose level for a T1DM patient. There already exist publications about modeling of insulin–glucose dynamics. They go from simple models [14] to very detailed ones, which are developed using compartmental techniques [43, 80, 124]. On the other hand, there are different publications about identification of glucose–insulin dynamics; most of them are based on linear approximation as AR (AutoRegresive) [125], ARX (AutoRegresive eXternal) [130, 75], ARMA (AutoRegresive Moving Average), or ARMAX (AutoRegresive Moving Average with eXogenous inputs) models [29]. However, due to the nonlinear behavior of insulin dynamics, a linear model is limited; hence, nonlinear identification techniques, like neural networks, are alternative options to determine mathematical models for glucose dynamics in T1DM [28]. The most used neural networks structures are feedforward networks and recurrent ones [3]. Since the seminal

paper [86], there has been a continuously increasing interest in applying neural net-

works to identification and control of nonlinear systems [109]. In addition, many of

the results known for conventional system identification are applicable to the neural

network based identification as well [71]. The most well-known training approach for

recurrent neural networks (RNN) is the back propagation through time [137]. How-

ever, it is a first order gradient descent method, and hence its learning speed could be

very slow [65]. Recently, Extended Kalman filter (EKF) based algorithms have been

introduced to train neural networks [3, 32]. With EKF based algorithms, the learning

convergence is improved [65]. EKF training of neural networks, both feedforward

and recurrent ones, has proven to be reliable for many applications over the past ten

years [32]. Regarding neural network modeling and control of glucose level, [128]

uses an artificial neural network (ANN) with a nonlinear model predictive control

technique; however, system identification is performed off-line using simulated data.

In contrast, the present chapter is based on on-line neural identification.

A number of different techniques have been applied for diabetes closed-loop con-

trol. These schemes include the classical PID control scheme [79, 55, 99, 126, 103],

the MPC (model predictive control) [55, 75, 102], and MPILC (a combination of

iterative learning control and MPC) [132], among others. All those control schemes

are exhaustively reviewed in [24, 101, 100], where advantages and disadvantages for

each one of them are explained. The PID scheme is the most used for control indus-

trial processes due to its simplicity and ease of tuning; for post-prandial effects of glucose control, it is difficult to ensure asymptotic zero error because it reacts slowly and induces hypoglycemia periods [24]. The MPC has been one of the most applied techniques; however, there is not a formal demonstration of its stability.

Recently, new publications report improved results for blood glucose level control. In [6] an intelligent on-line feedback strategy is presented for diabetic patients using adaptive critic neural networks. This technique is based on nonlinear optimal control theory; the authors perform a comparison with linear quadratic regulators (LQR), which illustrates how their technique is better in the sense that it never presents hypoglycemia and the necessity of negative control is eliminated. In [54], the authors apply a linear robust μ-synthesis technique to control T1DM; the controller obtained is applied to the novel Liu–Tang model [70], using a transformation to describe a T1DM patient; however, it does not consider hard constraints. An H_∞-based controller is used in [111] for T1DM, considering physiological effects due to exercise and nocturnal hypoglycemia, obtaining good results; however, robust H_∞ is necessary in order to prevent hypoglycemia events. According to these publications, optimal control is a technique widely studied for glucose level control problems.

7.2 PASSIVITY APPROACH

In this section, the inverse optimal trajectory tracking control scheme is applied to glycemic control of T1DM patients, by combining the results of Section 2.5.3 and Section 3.3.

Figure 7.2 shows a block diagram which portrays how the compartmental model proposed in [124], used as a virtual patient, is connected to the on-line neural identifier, and how the neural model is used to determine the control law; the objective of the control scheme is to develop an artificial pancreas, as depicted in Figure 7.3, which automatically provides to the patient an appropriate insulin dose; the block limited by dotted lines is used only in open loop, in order to simulate the virtual patient behavior. The virtual patient uses as inputs the total glucose absorbed by the patient gut with every meal and the insulin in the plasma calculated by the inverse optimal control law; then the on-line neural identifier captures the dynamics of the virtual patient. The model determined by the neural identifier is used to calculate the inverse optimal control law in order to obtain the insulin dose to be supplied to both the virtual patient and the neural model. The desired trajectory $(x_{\delta,k})$ is obtained by means of the model proposed by Lehmann and Deutsch [63]; constant K_m is used to obtain the reference for the normal range of glucose considering the postprandial effect. We select the trajectory to be tracked as the glucose level of a healthy person in order to improve the T1DM patient's well-being.

7.2.1 VIRTUAL PATIENT

The virtual patient plays an important role for the control law development; consequently it is important to select a model which represents a patient behavior as realistically as possible. There already exist very detailed models of insulin–glucose dynamics.They go from simple models [14] to very detailed ones [43, 80, 124]. Any one of the above could be selected for this work. However, the Sorensen model [124] is selected because it is the most complete one.

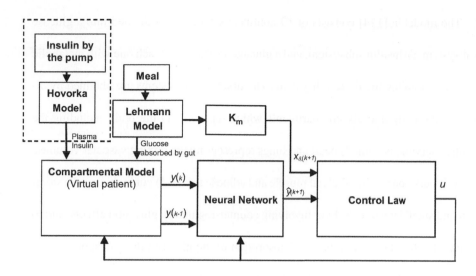

FIGURE 7.2   Closed loop diagram for the control law.

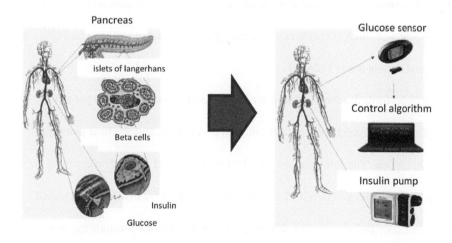

FIGURE 7.3 **(SEE COLOR INSERT)**   Artificial pancreas scheme. (From G. Quiroz, *Bio-*

signaling Weighting Functions to Handle Glycaemia in Type 1 Diabetes Mellitus via Feedback

Control System (in Spanish). PhD thesis, IPICYT, San Luis Potosi, Mexico, 2008. With per-

mission.)

The model in [124] consists of 19 nonlinear ODEs and is divided into a glucose subsystem, an insulin subsystem, and a glucagon subsystem, each one with metabolic rates for coupling the three subsystems.The first two subsystems are obtained by dividing the body into six compartments which represent 1) the brain including the central nervous system, 2) heart and lungs representing the mixing vascular volumes, 3) periphery concerning skeletal muscle and adipose tissue, 4) gut including the stomach and small intestine, 5) liver involving counter-regulatory glucagon effects, and 6) kidney for filtration, excretion, and absorption of the glucose; the glucagon subsystem is modeled as a single blood pool compartment. Figure 7.4 represents the scheme for the compartmental model of the glucose mass balance; the diagram includes the route for the control input and the glucose meal disturbance. The pancreas model is removed to represent the C-peptide negative endogenous response to glycemia [124]. Equations and details are in [124, 114].

In order to perform the required identification, the inputs to the virtual patient are the total glucose absorbed by the patient gut with every meal, and the time evolution for insulin in plasma. The total glucose absorbed by the patient gut with every meal is calculated by the model given in [63], and the time evolution for insulin is calculated by the control law; before entering the virtual patient, it passes through the model proposed in [43]. The output of the virtual patient is the glucose in the periphery interstitial fluid space.

Simulation results are presented for two different patients in order to see how the control law is able to control blood glucose levels, avoiding hyperglycemia and

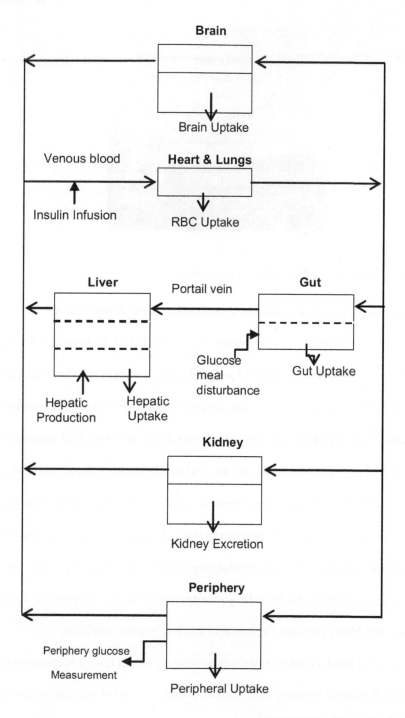

FIGURE 7.4   Schematic representation of the compartmental model for glucose mass balance.

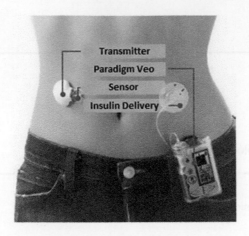

FIGURE 7.5 **(SEE COLOR INSERT)**   Insulin pump.

hypoglycemia. The experimental data correspond to two different patients. The first

one is a female patient 23 years old, with a 14-year-old T1DM diagnosis, 1.68 m and

58.5 kg. The second patient is a male 14 years old, with a 13-year-old T1DM diagnosis,

1.55 m and 54 kg. Patients do not present any other complication or disease associated

with diabetes mellitus. Data are collected for normal days, under medical supervision,

with standard ingesta (three meals per day and snacks between them) and without

exercise events. The patients have to register the quantity of carbohydrates every time

they take a meal, and they have to program an insulin bolus to the Paradigm® real-time

insulin pump[1] (see Figure 7.5). The bolus programmed by patients always occurs at

the same time as ingesta. Additionally, patients can program corrector boluses for the

case that their blood glucose level is near to a hyperglycemic condition.

   On the other hand, in order to simulate the virtual patient [124], it is convenient to

consider a dynamical model of subcutaneous insulin absorption and glucose absorp-

---

[1] Trademark of Medtronic MiniMed, Inc. Bayer.

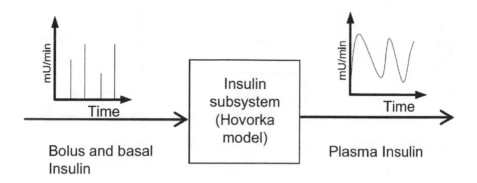

FIGURE 7.6   Diagram for the subsystem of the Hovorka model.

tion after oral ingestion. That is, insulin is infused to a patient in two ways; the basal

rate is provided continuously and boluses are given to correct meals. Both of these

are provided by the subcutaneous route; it is then convenient to obtain plasma insulin

concentration after absorption, as Figure 7.6 illustrates. To this end, the insulin sub-

system of the mathematical model proposed in [43] is considered, which comprises

three ordinary differential equations ODE.

The mathematical model proposed in [63] is employed in order to approximate

continuous glucose absorption by the gut, after carbohydrate oral consumption, as

shown in Figure 7.7. The outputs which are obtained with the Hovorka [43] and

Lehmann and Deutsch [63] models are used as inputs to the Sorensen model. The

output is the peripherial interstitial glucose. Figure 7.8 presents the inputs and the

outputs of the Sorensen model which are obtained for the female patient. Simulations

are implemented using MATLAB.

Results for the virtual patient simulation are presented in Figure 7.8 (female patient)

and Figure 7.9 (male patient). It can be seen that the patients have hypoglycemia and

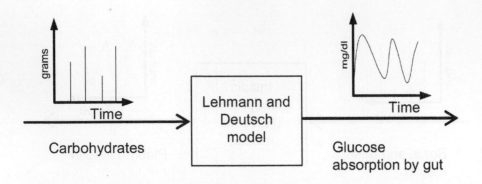

FIGURE 7.7   Diagram for the Lehmann and Deutsch model.

hyperglycemia periods, although the insulin pump is delivering insulin.

7.2.2 STATE SPACE REPRESENTATION

For determination of the optimal regressor structure, the Cao methodology [19] is used. This method determines the minimum embedding dimension. It has the following advantages: (1) it does not contain any subjective parameters except for the time-delay embedding; (2) it does not strongly depend on how many data points are available; (3) it can clearly distinguish deterministic signals from stochastic signals; (4) it works well for time series from high-dimensional attractors; and (5) it is computationally efficient. The lagged recurrent inputs to the RMLP are equal to $k = 2$.

To synthesize a control law based on the virtual patient [124] would be a difficult task. Therefore, a neural model is obtained from the virtual patient in order to implement the control law. The identification of the virtual patient is done with neural network identification as explained in Section 2; then (2.48) becomes

$$\hat{y}_{k+1} = \sum_{i=0}^{5} w_{1i}^{(2)} v_i \text{ with } v_0 = +1 \tag{7.1}$$

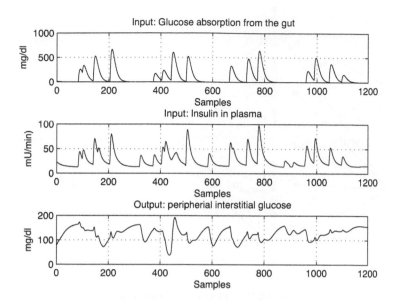

FIGURE 7.8   Inputs and output for the Sorensen model (female patient).

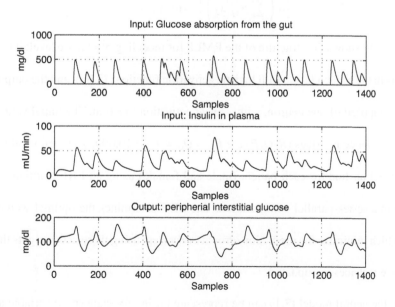

FIGURE 7.9   Inputs and output for the Sorensen model (male patient).

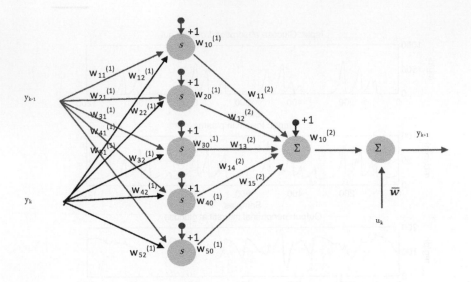

FIGURE 7.10   Structure of the RMLP for glucose level modeling.

where

$$v_i = \left[S \left(\sum_{j=0}^{2} w_{ij}^{(1)} x_j \right) \right] \text{ with } x_0 = +1.$$

Figure 7.10 shows the structure of the RMLP for modeling of glucose level; it has

5 neurons in the hidden layer, with logistic activation functions (2.36), and the output

layer is composed of one neuron, with a linear activation function. The initial values

for the covariance matrices (R,Q,P) are $R_0 = Q_0 = P_0 = 10{,}000$. The identification

is performed on-line for each patient (male and female) using an EKF learning al-

gorithm in a series-parallel configuration. The EKF determines the optimal weight

values which minimize the prediction error at every step; using these new weights the

covariance matrices are updated.

Then, the neural model (7.1) can be represented using the state space variable as follows:

$$x_{1,k+1} = f_1(x_k) \tag{7.2}$$

$$x_{2,k+1} = f_2(x_k) + g(x_k)u(x_k) \tag{7.3}$$

$$\hat{y}_k = x_{2,k}$$

$$y_k = h(x_k, x_{\delta,k+1}) + J(x_k)u_k$$

$$f(x_k) = \begin{bmatrix} f_{1,k} \\ f_{2,k} \end{bmatrix} = \begin{bmatrix} x_{2,k} \\ \hat{y}_{k+1} \end{bmatrix}$$

$$g(x_k) = \begin{bmatrix} 0 \\ w' \end{bmatrix}$$

where $h(x_k, x_{\delta,k+1})$ is equal to (3.44) and $J(x_k)$ is equal to (3.45), $x_{2,k+1}$ is the glucose level, and u_k is the insulin dose.

Results for the neural identification are presented in Figure 7.11 for the female patient and Figure 7.12 for the male patient which correspond to the neural network identifier and the identification error. The sample time is 5 minutes because the Paradigm real-time continuous glucose monitoring system takes a sample every 5 minutes. The total number of samples is 1200 for the female patient, equivalent to 4 days of monitoring, and 1400 for the male patient, equivalent to almost 5 days of monitoring.

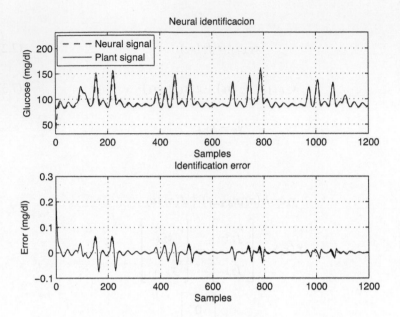

FIGURE 7.11   Neural identification and the identification error (female patient).

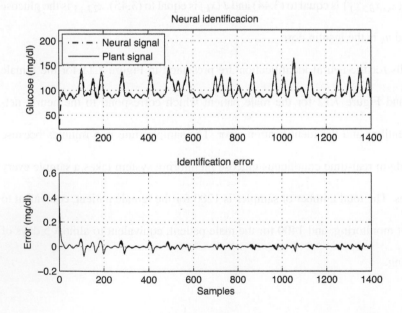

FIGURE 7.12   Neural identification and the identification error (male patient).

7.2.3 CONTROL LAW IMPLEMENTATION

In order to determine $\overline{P} = K_1^T P K_1$, P_1 and K_1 are selected by simulation so as to fulfill

(3.42) conditions, as follows:

$$
P = \begin{bmatrix} 14 & 11.5 \\ 11.5 & 14 \end{bmatrix} \text{ and } K_1 = \begin{bmatrix} 46 & 0 \\ 0 & 6.5 \end{bmatrix}.
$$

The desired trajectory $(x_{\delta,k})$ is obtained using the model proposed by Lehmann and

Deutsch [63] in order to take into account the post-prandial effect. It would be easier

if the glucose remains within a band, but the desired trajectory is determined in such a

way as to perform more realistic experiments. Forcing the patient blood glucose level

to track the corresponding trajectory of a healthy person improves the well being of the

patient. This statement is corroborated by the complete elimination of hypoglycemia,

which is in contrast with regulation type control of glucose level. In order to achieve

trajectory tracking, the control law (3.60) is implemented. The tracking performance

and the tracking error are displayed in Figure 7.13 for the female patient and for the

male patient in Figure 7.14. Figure 7.15 presents the difference between the insulin

supplied to the patient with the insulin pump (open loop) and the insulin which is

calculated by the proposed control law for the female patient and Figure 7.16 for the

male patient. For the female patient the mean of the insulin supplied by the pump

is 27.51 mU/min and for the male patient 25.47 mU/min; the mean of the insulin

calculated by the proposed control law for the female patient is 25 mU/min and for

the male patient 26.81mU/min. Figure 7.15 also displays the difference between the

glucose in the plasma taken from the patient with the continuous glucose monitoring

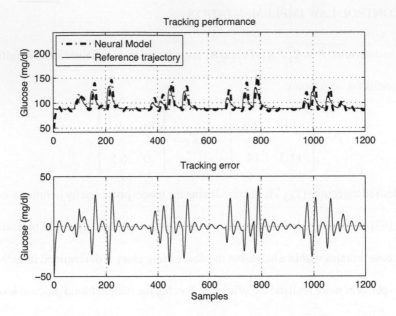

FIGURE 7.13   Tracking performance of glucose in plasma from a patient with T1DM (female

patient).

system by MiniMed Inc. (open-loop) and the glucose in the plasma with the proposed

control law for the female patient and Figure 7.16 for the male patient. It can be noticed

that the glucose reaches values above the reference, due to the carbohydrates ingested

by the patient, which acts like a perturbation; however, the control law corrects the

glucose level through insulin supply. Finally, Figure 7.17 displays the cost functional

evaluation.

7.3   CLF APPROACH

This section combines the results of Section 2.5.3 and Section 4.4 to achieve trajectory

tracking for glycemic control of T1DM patients.

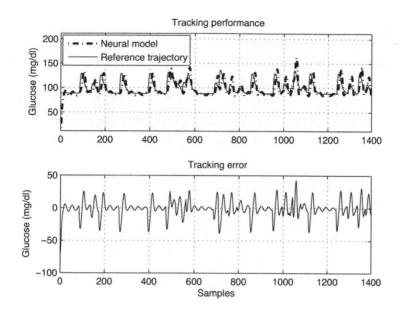

FIGURE 7.14   Tracking performance of glucose in plasma from a patient with T1DM (male

patient).

For the implementation of control law (4.77), the values for P and R are selected

as

$$P = \begin{bmatrix} 128 & 198.4 \\ 198.4 & 128 \end{bmatrix}$$

and $R = 0.48$.

The desired trajectory $x_{\delta,k}$ is obtained using the model proposed by Lehmann

and Deutsch [63] in order to take into account the post-prandial effect. Forcing the

patient blood glucose level to track the corresponding trajectory of a healthy person

improves the well being of the patient. This statement is corroborated by the complete

elimination of hypoglycemia, which is in contrast with regulation type control of

glucose level.

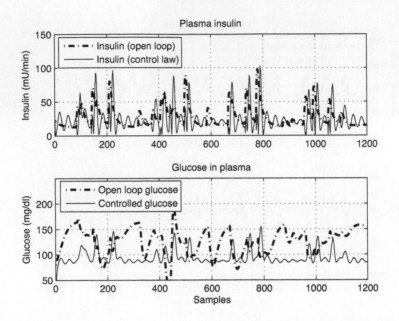

FIGURE 7.15   Control law for controlling the periods of hyperglycemia and hypoglycemia

in a T1DM patient (female patient).

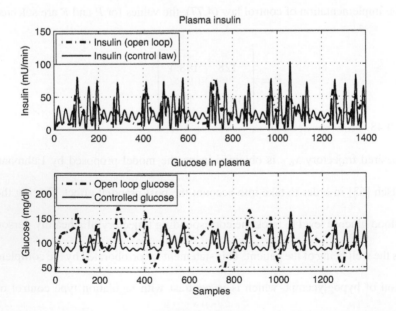

FIGURE 7.16   Control law for controlling the periods of hyperglycemia and hypoglycemia

in a T1DM patient (male patient).

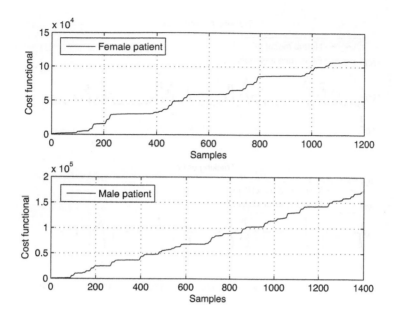

FIGURE 7.17   Cost functional evaluation for both patients.

7.3.1   SIMULATION RESULTS VIA CLF

The tracking performance and the tracking error are displayed in Figure 7.18 for

the female patient, and for the male patient in Figure 7.19. Figure 7.20 presents the

difference between the insulin supplied to the patient with the insulin pump (open

loop) and the insulin which is calculated by the proposed control law for the female

patient and Figure 7.21 for the male patient. For the female patient the mean of the

insulin supplied by the pump is 27.51 mU/min and for the male patient it is 25.47

mU/min; the mean of the insulin calculated by the proposed control law for the

female patient is 25.91 mU/min and for the male patient 25.81 mU/min. Figure 7.20

also displays the difference between the glucose in the plasma taken from the patient

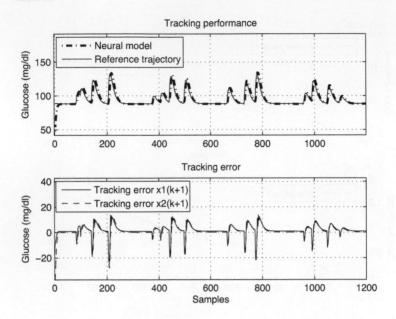

FIGURE 7.18   Tracking performance of glucose in plasma for the female patient with T1DM.

with the continuous glucose monitoring system by MiniMed Inc. (open-loop) and

the glucose in the plasma with the proposed control law for the female patient and

Figure 7.21 for the male patient. It can be noticed that the glucose reaches values

above the reference, due to the carbohydrates ingested by the patient, which acts like

a perturbation; however, the control law corrects the glucose level through insulin

supply.

7.3.2   PASSIVITY VERSUS CLF

This section presents a brief comparison between the inverse optimal neural control

based on the passivity approach (Section 7.2) and the CLF approach (Section 7.3).

Figure 7.22 shows simulation results for the male patient using control via passivity

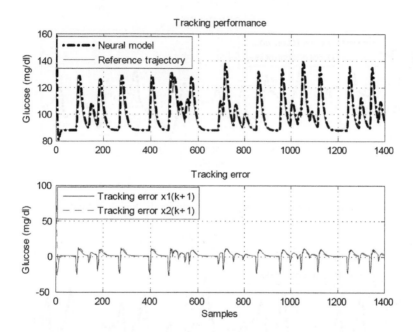

FIGURE 7.19   Tracking performance of glucose in plasma for the male patient with T1DM.

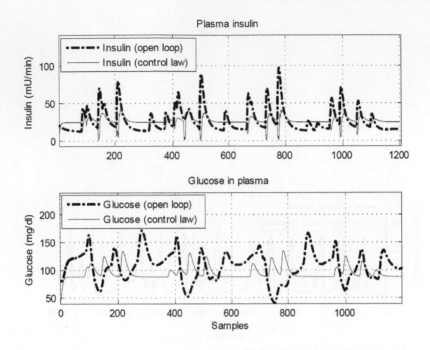

FIGURE 7.20   Open-loop and closed loop comparison for the female patient.

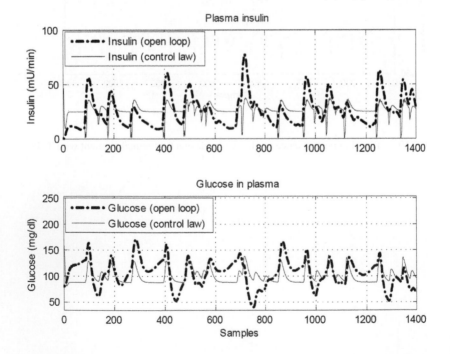

FIGURE 7.21   Open loop and closed loop comparison for the male patient.

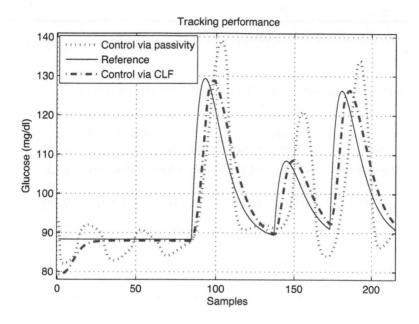

FIGURE 7.22   Comparison between the control scheme via passivity and via CLF for the male patient.

and control via CLF.

Table 7.1 displays a statistical comparison between control via CLF and control via passivity for female (F) and male (M) patient; where it can be seen that the control scheme via CLF is more robust against perturbations, in this case perturbations are the meal periods. Table 7.1 illustrates that glucose control via CLF presents advantages over glucose control via passivity, in order to achieve trajectory tracking.

7.4   CONCLUSIONS

This chapter proposes the use of recurrent neural networks (RNN) for modeling and control of glucose–insulin dynamics in T1DM patients. The proposed RNN used in

TABLE 7.1

Comparison between passivity and CLF approaches for glucose control.

	Mean square error			Standard deviation		
	Open-loop	Passivity	CLF	Open-loop	Passivity	CLF
F	1782.2	0.0026	0.0018	29.07	0.057	0.039
M	665.00	0.0033	0.0013	25.00	0.058	0.036

our experiments captures very well the complexity associated with blood glucose level for type 1 diabetes mellitus patients. The proposed RNN is used to derive an affine dynamical mathematical model for type 1 diabetes mellitus. The affine mathematical model is obtained with the aim of applying inverse optimal neural control. Once the neural model is obtained, it is used to synthesize the inverse optimal controller. The proposed scheme stabilizes the system along a desired trajectory and simulation results illustrate the applicability of the proposed scheme. Indeed, this scheme greatly improves the regulation of the blood glucose level in T1DM patients, increasing slightly the insulin quantity. The proposed control law is capable of eliminating hyperglycemia and hypoglycemia for patients. This preliminary work gives the pattern to develop a computing environment to begin clinical trials.

8 Conclusions

This book proposes a novel discrete-time inverse optimal control scheme, which achieves stabilization and trajectory tracking for nonlinear systems and is inverse optimal in the sense that, a posteriori, it minimizes a cost functional. To avoid the Hamilton–Jacobi–Bellman equation solution, we proposed a discrete-time quadratic control Lyapunov function (CLF). The controller synthesis is based on two approaches: *a*) inverse optimal control based on passivity, in which the storage function is used as a CLF, and *b*) an inverse optimal control based on the selection of a CLF. Furthermore, a robust inverse optimal control is established in order to guarantee stability for nonlinear systems, which are affected by internal and/or external disturbances.

We use discrete-time recurrent neural networks to model uncertain nonlinear systems; thus, an explicit knowledge of the plant is not necessary. The proposed approach is successfully applied to implement a robust controller based on a recurrent high order neural network identifier and inverse optimality. By means of simulations, it can be seen that the required goal is achieved, i.e., the proposed controller maintains stability of the plant with unknown parameters. For neural network training, an on-line extended Kalman filter is performed.

The applicability of the proposed controllers is illustrated, via simulations, by stabilization and trajectory tracking of academic and physical systems. Additionally, the control scheme is applied to the control of T1DM.

The major contributions of this book are

- To synthesize an *inverse optimal discrete-time controller for nonlinear systems via passivity (Chapter 3)*. The controller synthesis is based on the selection of a storage function, which is used as a CLF candidate, and a passifying law to render the system passive. This controller achieves stabilization and trajectory tracking for discrete-time nonlinear systems and is inverse optimal in the sense that, a posteriori, it minimizes a cost functional.

- To synthesize an *inverse optimal discrete-time controller for nonlinear systems via a CLF (Chapter 4 and Chapter 5)*. A CLF candidate for the obtained control law is proposed such that stabilization and trajectory tracking for discrete-time nonlinear systems are achieved; a posteriori, a cost functional is minimized. The CLF depends on a fixed parameter in order to satisfy stability and optimality conditions. A posteriori, the speed gradient algorithm is utilized to compute this CLF parameter.

- To establish a *neural inverse optimal discrete-time control scheme for uncertain nonlinear systems (Chapter 6)*. For this neural scheme, an assumed uncertain discrete-time nonlinear system is identified by a RHONN model, which is used to synthesize the inverse optimal controller in order to achieve stabilization and trajectory tracking. The neural learning is performed on-line through an extended Kalman filter.

- To develop a methodology to deal with glycemic control of T1DM patients (Chapter 7). The proposed control law calculates the adequate insulin delivery rate in order to prevent hyperglycemia and hypoglycemia levels in T1DM patients. For this control methodology, an on-line neural identifier is proposed to model type 1 diabetes mellitus virtual patient behavior, and finally the inverse optimal control scheme is applied for the neural identifier model to achieve tracking along a desired reference: the glucose absorption of a healthy person.

References

1. T. Ahmed-Ali, F. Mazenc, and F. Lamnabhi-Lagarrigue. Disturbance attenuation for discrete-time feedforward nonlinear systems. *Lecture Notes in Control and Information Sciences*, 246:1–17, 1999.

2. A. Al-Tamimi and F. L. Lewis. Discrete-time nonlinear HJB solution using approximate dynamic programming: Convergence proof. *IEEE Transactions on Systems, Man, Cybernetics—Part B*, 38(4):943–949, 2008.

3. A. Y. Alanis, E. N. Sanchez, and A. G. Loukianov. Discrete time adaptive backstepping nonlinear control via high order neural networks. *IEEE Transactions on Neural Networks*, 18(4):1185–1195, 2007.

4. A. Y. Alanis, E. N. Sanchez, and A. G. Loukianov. Discrete-time backstepping synchronous generator stabilization using a neural observer. In *Proceedings of the 17th IFAC World Congress*, pages 15897–15902, Seoul, Korea, 2008.

5. A. Y. Alanis, E. N. Sanchez, A. G. Loukianov, and M. A. Perez. Discrete-time output trajectory tracking by recurrent high-order neural network control. *IEEE Transactions on Control System Technology*, 18(1):11–21, 2010.

6. S. F. Ali and R. Padhi. Optimal blood glucose regulation of diabetic patients using single network adaptive critics. *Optimal Control Applications and Methods*, 32:196–214, 2009.

7. G. L. Amicucci, S. Monaco, and D. Normand-Cyrot. Control Lyapunov sta-

bilization of affine discrete-time systems. In *Proceedings of the 36th IEEE Conference on Decision and Control*, volume 1, pages 923–924, San Diego, CA, USA, Dec 1997.

8. B. D. O. Anderson and J. B. Moore. *Optimal Control: Linear Quadratic Methods*. Prentice-Hall, Englewood Cliffs, NJ, USA, 1990.

9. M. A. Arjona, R. Escarela-Perez, G. Espinosa-Perez, and J. Alvarez-Ramirez. Validity testing of third-order nonlinear models for synchronous generators. *Electric Power Systems Research*, (79):953–958, 2009.

10. Z. Artstein. Stabilization with relaxed controls. *Nonlinear Analysis: Theory, Methods and Applications*, 7(11):1163–1173, 1983.

11. T. Basar and G. J. Olsder. *Dynamic Noncooperative Game Theory*. Academic Press, New York, USA, 2nd edition, 1995.

12. R. E. Bellman. *Dynamic Programming*. Princeton University Press, Princeton, NJ, USA, June 1957.

13. R. E. Bellman and S. E. Dreyfus. *Applied Dynamic Programming*. Princeton University Press, Princeton, NJ, USA, 1962.

14. R. N. Bergman, Y. Z. Ider, C. R. Bowden, and C. Cobelli. Quantitative estimation of insulin sensitivity. *Am. J. Physiology, Endocrinology and Metabolism*, 235:667–677, 1979.

15. A. Berman, M. Neumann, and R. J. Stem. *Nonnegative Matrices in Dynamic Systems*. Wiley, New York, USA, 1989.

16. B. W. Bode. *Medical Management of Type 1 Diabetes, 4th ed.* American

Diabetes Association, Alexandria, VA, USA, 1998.

17. B. Brogliato, R. Lozano, B. Maschke, and O. Egeland. *Dissipative Systems Analysis and Control: Theory and Applications*. Springer-Verlag, Berlin, Germany, 2nd edition, 2007.

18. C. I. Byrnes and W. Lin. Losslessness, feedback equivalence, and the global stabilization of discrete-time nonlinear systems. *IEEE Transactions on Automatic Control*, 39(1):83–98, 1994.

19. L. Cao. Practical method for determining the minimum embedding dimension of a scalar time series. *Physica D: Nonlinear Phenomena*, 110(1):43–50, 1997.

20. C. E. Castaneda, A. G. Loukianov, E. N. Sanchez, and B. Castillo-Toledo. Discrete-time neural sliding-mode block control for a dc motor with controlled flux. *IEEE Transactions on Industrial Electronics*, 59(2):1194–1207, 2012.

21. J. Casti. On the general inverse problem of optimal control theory. *Journal of Optimization Theory and Applications*, 32(4):491–497, 1980.

22. B. Castillo-Toledo, S. D. Gennaro, A. G. Loukianov, and J. Rivera. Discrete time sliding mode control with application to induction motors. *Automatica*, 44(12):3036–3045, 2008.

23. B. Castillo-Toledo, S. D. Gennaro, A. G. Loukianov, and J. Rivera. Discrete time sliding mode control with application to induction motors. *Automatica*, 44(12):3036–3045, 2008.

24. C. Cobelli, C. D. Man, G. Sparacio, L. Magni, G. De Nicolao, and B. P. Kovatchev. Diabetes: models, signals and control. *IEEE Reviews in Biomedical*

Engineering, 2:54–94, 2009.

25. J. J. Craig. *Introduction to Robotics: Mechanics and Control.* Addison Wesley Longman, USA, 1989.

26. C. Cruz-Hernandez, J. Alvarez-Gallegos, and R. Castro-Linares. Stability of discrete nonlinear systems under nonvanishing perturbations: application to a nonlinear model–matching problem. *IMA Journal of Mathematical Control & Information*, 16:23–41, 1999.

27. J. W. Curtis and R. W. Beard. A complete parameterization of CLF-based input-to-state stabilizing control laws. *International Journal of Robust and Nonlinear Control*, 14:1393–1420, 2004.

28. A. K. El-Jabali. Neural network modeling and control of type 1 diabetes mellitus. *Bioprocess and Biosystems Engineering*, 27(2):75–79, 2005.

29. M. Eren-Oruklu, A. Cinar, L. Quinn, and D. Smith. Adaptive control strategy for regulation of blood glucose levels for patients with type 1 diabetes. *Journal of Process Control*, 19:1333–1346, 2009.

30. G. Escobar, R. Ortega, H. Sira-Ramirez, J. P. Vilian, and I. Zein. An experimental comparison of several non-linear controllers for power converters. *IEEE Control Systems Magazine*, 19(1):66–82, 1999.

31. L. Farina and S. Rinaldi. *Positive Linear Systems: Theory and Applications.* Wiley, New York, USA, 2000.

32. L. A. Feldkamp, D. V. Prokhorov, and T. M. Feldkamp. Simple and conditioned adaptive behavior from Kalman filter trained recurrent networks. *Neural*

Networks, 16:683–689, 2003.

33. A. L. Fradkov and A. Y. Pogromsky. *Introduction to Control of Oscillations and Chaos*. World Scientific Publishing Co., Singapore, 1998.

34. R. A. Freeman and P. V. Kokotović. Optimal nonlinear controllers for feedback linearizable systems. In *Proceedings of the 14th American Control Conference, 1995*, volume 4, pages 2722–2726, Seattle, WA, USA, June 1995.

35. R. A. Freeman and P. V. Kokotović. Inverse optimality in robust stabilization. *SIAM Journal on Control and Optimization*, 34(4):1365–1391, 1996.

36. R. A. Freeman and P. V. Kokotović. *Robust Nonlinear Control Design: State-Space and Lyapunov Techniques*. Birkhauser Boston Inc., Cambridge, MA, USA, 1996.

37. R. A. Freeman and J. A. Primbs. Control Lyapunov functions: New ideas from an old source. In *Proceedings of the 35th IEEE Conference on Decision and Control*, pages 3926–3931, Kobe, Japan, Dec 1996.

38. R. Gourdeau. Object-oriented programming for robotic manipulator simulation. *IEEE Robotics and Automation*, 4(3):21–29, 1997.

39. J. W. Grizzle, M. D. Benedetto, and L. Lamnabhi-Lagarrigue. Necessary conditions for asymptotic tracking in nonlinear systems. *IEEE Transactions on Automatic Control*, 39(9):1782–1794, 1994.

40. R. Grover and P. Y. C. Hwang. *Introduction to Random Signals and Applied Kalman Filtering, 2nd ed.* John Wiley and Sons, New York, USA, 1992.

41. W. M. Haddad, V. Chellaboina, J. L. Fausz, and C. Abdallah. Optimal discrete-

time control for non-linear cascade systems. *Journal of the Franklin Institute*, 335(5):827–839, 1998.

42. S. Haykin. *Kalman Filtering and Neural Networks*. Wiley, Upper Saddle River, NJ, USA, 2001.

43. R. Hovorka, V. Canonico, L. J. Chassin, U. Haueter, M. Massi-Benedetti, M. Orsini-Federici, T. R. Pieber, H. Schaller, L. Schaupp, T. Vering, and M. Wilinska. Non-linear model predictive control of glucose concentration in subjects with type 1 diabetes. *Physiological Measurement*, 25:905–920, 2004.

44. A. Isidori. *Nonlinear Control Systems*. Springer-Verlag, London, UK, 1997.

45. T. Kaczorek. *Positive 1-D and 2-D Systems*. Springer-Verlag, Berlin, Germany, 2002.

46. R. E. Kalman. When is a linear control system optimal? *Transactions of the ASME, Journal of Basic Engineering, Series D*, 86:81–90, 1964.

47. J. G. Kassakian, M. F. Schlecht, and G. C. Verghese. *Principles of Power Electronics*. Adison Wesley, MA, USA, 1991.

48. C. M. Kellett and A. R. Teel. Results on discrete-time control-Lyapunov functions. In *Proceedings of the 42nd IEEE Conference on Decision and Control, 2003*, volume 6, pages 5961–5966, Maui, Hawaii, USA, Dec 2003.

49. H. K. Khalil. *Nonlinear Systems*. Prentice-Hall, Upper Saddle River, NJ, USA, 1996.

50. Y. H. Kim and F. L. Lewis. *High-Level Feedback Control with Neural Networks*. World Scientific, Singapore, 1998.

51. D. E. Kirk. *Optimal Control Theory: An Introduction*. Prentice-Hall, Englewood Cliffs, NJ, USA, 1970.

52. P. Kokotović and M. Arcak. Constructive nonlinear control: Progress in the 90's. Technical Report CCEC98-1209, 1998. Extended Text of the Plenary Talk, 14th Triennial World Congress, Beijing, China, July 5-9, 1999.

53. E. B. Kosmatopoulos, M. M. Polycarpou, M. A. Christodoulou, and P. A. Ioannou. High-order neural network structures for identification of dynamical systems. *IEEE Transactions on Neural Networks*, 6(2):422–431, 1995.

54. L. Kovács, B. Kulcsár, A. Gyórgy, and Z. Benyó. Robust servo control of a novel type 1 diabetic model. *Optimal Control Applications and Methods*, 32:215–238, 2010.

55. B. P. Kovatchev, M. Breton, C. D. Man, and C. Cobelli. In silico preclinical trials: a proof concept in closed-loop control of type 1 diabetes. *Journal of Diabetes Science and Technology*, 3(3):44–55, 2009.

56. M. Krstić and H. Deng. *Stabilization of Nonlinear Uncertain Systems*. Springer-Verlag, Berlin, Germany, 1998.

57. M. Krstić and Z. Li. Inverse optimal design of input-to-state stabilizing nonlinear controllers. *IEEE Transactions on Automatic Control*, 43(3):336–350, 1998.

58. M. Krstić and Z. Li. Inverse optimal design of input-to-state stabilizing nonlinear controllers. *IEEE Transactions on Automatic Control*, 43(3):336–350, 1998.

59. D. S. Laila and A. Astolfi. Discrete-time IDA-PBC design for underactuated Hamiltonian control systems. In *Proceedings of the 2006 American Control*

Conference, pages 188–193, Minneapolis, MN, USA, June 2006.

60. C. Lalli, M. Ciofetta, P. D. Sindaco, E. Torlone, S. Pampanelli, P. Compagnucci, M. G. Cartechini, L. Bartocci, P. Brunetti, and G. B. Bolli. Long-term intensive treatment of type 1 diabetes with the short-acting insulin analog lispro in variable combination with nph insulin at meal-time. *Diabetes Care*, 22:468–477, 1999.

61. J. P. LaSalle. *The Stability and Control of Discrete Processes*. Springer-Verlag, Berlin, Germany, 1986.

62. P. D. Leenheer and D. Aeyels. Stabilization of positive linear systems. *Systems and Control Letters*, 44(4):259–271, 2001.

63. E. D. Lehmann and T. Deutsch. A physiological model of glucose-insulin interaction in type 1 diabetes mellitus. *Journal on Biomedical Engineering*, 14:235–242, 1992.

64. J. De Leon-Morales, O. Huerta-Guevara, L. Dugard, and J. M. Dion. Discrete-time nonlinear control scheme for synchronous generator. In *Proceedings of the 42nd Conference on Decision and Control*, pages 5897–5902, Maui, Hawaii, USA, Dec 2003.

65. C. Leunga and L. Chan. Dual extended Kalman filtering in recurrent neural networks. *Neural Networks*, 16:223–239, 2003.

66. F. L. Lewis, S. Jagannathan, and A. Yesildirek. *Neural Network Control of Robot Manipulators and Nonlinear Systems*. Taylor and Francis, London, UK, 1999.

67. F. L. Lewis and V. L. Syrmos. *Optimal Control*. John Wiley & Sons, New York,

USA, 1995.

68. W. Lin and C. I. Byrnes. Design of discrete-time nonlinear control systems via smooth feedback. *IEEE Transactions on Automatic Control*, 39(11):2340–2346, 1994.

69. W. Lin and C. I. Byrnes. Passivity and absolute stabilization of a class of discrete-time nonlinear systems. *Automatica*, 31(2):263–267, 1995.

70. W. Liu and F. Tang. Modeling a simplified regulatory system of blood glucose at molecular levels. *Journal of Theoretical Biology*, 252:608–620, 2008.

71. L. Ljung. *System Identification: Theory for the User, 2nd ed.* Prentice Hall, Upper Saddle River, NJ, USA, 1999.

72. A. G. Loukianov. Nonlinear block control with sliding modes. *Automation and Remote Control*, 57(7):916–933, 1998.

73. A. G. Loukianov and V. I. Utkin. Methods of reducing equations for dynamic systems to a regular form. *Automation and Remote Control*, 42(4):413–420, 1981.

74. Y. Maeda. Euler's discretization revisited. In *Proceedings of the Japan Academy*, volume 71, pages 58–61, 1995.

75. L. Magni, M. Forgione, C. Toffanin, C. D. Man, B. Kovatchev, G. De Nicolao, and C. Cobelli. Run-to-run tuning of model predictive control for type 1 diabetes subjects, in silico trial. *Journal of Diabetes Science and Technology*, 3:1091–1098, 2009.

76. L. Magni and R. Scattolini. *Assessment and Future Directions of Nonlinear*

Model Predictive Control, volume 358 of *Lecture Notes in Control and Information Sciences*. Springer-Verlag, Berlin, Germany, 2007.

77. L. Magni and R. Sepulchre. Stability margins of nonlinear receding horizon control via inverse optimality. *Systems and Control Letters*, 32(4):241–245, 1997.

78. L. Mailleret. *Positive Control for a Class of Nonlinear Positive Systems*. In *Lecture Notes in Control and Information Sciences*. Springer-Verlag, Berlin, Germany, 2003.

79. C. D. Man, D. M. Raimondo, R. A. Rizza, and C. Cobelli. Gim simulation software of meal glucose-insulin model. *Journal of Diabetes Science and Technology*, 1(3):323–330, 2007.

80. C. D. Man, R. A. Rizza, and C. Cobelli. Meal simulation model of the glucose-insulin system. *IEEE Transactions on Biomedical Engineering*, 54(10):1740–1749, 2007.

81. M. Margaliot and G. Langholz. Some nonlinear optimal control problems with closed-form solutions. *International Journal of Robust and Nonlinear Control*, 11(14):1365–1374, 2001.

82. R. Marino and P. Tomei. *Nonlinear Control Design: Geometric, Adaptive and Robust*. Prentice Hall, Hertfordshire, UK, 1996.

83. S. Monaco and D. Normand-Cyrot. Nonlinear representations and passivity conditions in discrete time. *Lecture Notes in Control and Information Sciences*, 245:422–432, 1999.

84. P. J. Moylan. Implications of passivity in a class of nonlinear systems. *IEEE Transactions on Automatic Control*, 19(4):373–381, 1974.

85. P. J. Moylan and B. D. O. Anderson. Nonlinear regulator theory and an inverse optimal control problem. *IEEE Transactions on Automatic Control*, 18(5):460–465, 1973.

86. K. S. Narendra and K. Parthasarathy. Identification and control of dynamical systems using neural networks. *IEEE Transactions on Neural Networks*, 1:4–27, 1990.

87. E. M. Navarro-López. *Dissipativity and passivity-related properties in nonlinear discrete-time systems*. PhD thesis, Universidat Politécnica de Catalunya, Barcelona, Spain, June 2002.

88. E. M. Navarro-López. Local feedback passivation of nonlinear discrete-time systems through the speed-gradient algorithm. *Automatica*, 43(7):1302–1306, 2007.

89. D. Nesic, A. R. Teel, and E. D. Sontag. Sufficient conditions for stabilization of sampled-data nonlinear systems via discrete-time approximations. *Systems & Control Letters*, 38(4–5):259–270, 1999.

90. M. Norgaard, N. K. Poulsen, and O. Ravn. Advances in derivative-free state estimation for nonlinear systems. Technical Report IMM-REP-1988-15 (revised edition), 1993.

91. T. Ohsawa, A. M. Bloch, and M. Leok. Discrete Hamilton-Jacobi theory and discrete optimal control. In *Proceedings of the 49th IEEE Conference on Decision*

and Control (CDC), pages 5438–5443, Dec 2010.

92. F. Ornelas, A. G. Loukianov, E. N. Sanchez, and E. J. Bayro-Corrochano. Planar robot robust decentralized neural control. In *Proceedings of the IEEE Multiconference on Systems and Control (MSC 2008)*, San Antonio, Texas, USA, Sep 2008.

93. F. Ornelas, E. N. Sanchez, and A. G. Loukianov. Discrete-time inverse optimal control for nonlinear systems trajectory tracking. In *Proceedings of the 49th IEEE Conference on Decision and Control*, pages 4813–4818, Atlanta, Georgia, USA, Dec 2010.

94. F. Ornelas-Tellez, A. G. Loukianov, E. N. Sanchez, and E. J. Bayro-Corrochano. Decentralized neural identification and control for uncertain nonlinear systems: Application to planar robot. *Journal of the Franklin Institute*, 347(6):1015–1034, 2010.

95. F. Ornelas-Tellez, E. N. Sanchez, and A. G. Loukianov. Discrete-time robust inverse optimal control for a class of nonlinear systems. In *Proceedings of the 18th IFAC World Congress*, pages 8595–8600, Milano, Italy, 2011.

96. R. Ortega, A. Loría, P. J. Nicklasson, and H. Sira-Ramirez. *Passivity-based Control of Euler-Lagrange Systems: Mechanical, Electrical and Electromechanical Applications*. Springer-Verlag, Berlin, Germany, 1998.

97. R. Ortega and M. W. Spong. Adaptive motion control of rigid robots: a tutorial. *Automatica*, 25(6):877–888, 1989.

98. R. Ortega, A. J. Van Der Schaft, I. Mareels, and B. Maschke. Putting energy

back in control. *IEEE Control Systems Magazine*, 21(2):18–33, 2001.

99. C. C. Palerm. Physiologic insulin delivery with insulin feedback: A control systems perspective. *Computer Methods and Programs in Biomedicine*, 102(2):130–137, 2011.

100. R. S. Parker and F. J. Doyle. Control-relevant modeling in drug delivery. *Advanced Drug Delivery Reviews*, 48:211–228, 2001.

101. R. S. Parker., F. J. Doyle, and N. A. Peppas. The intravenous route to blood glucose control, a review of control algorithms for non invasive monitoring and regulation in type 1 diabetic patients. *IEEE Engineering in Medicine and Biology Magazine*, 20(1):65–73, 2001.

102. S. D. Patek, W. Bequette, M. Breton, B. A. Buckingham, E. Dassau, F. J. Doyle, J. Lum, L. Magni, and H. Zizzcr. In silico preclinical trials: methodology and engineering guide to closed-loop control in type 1 diabetes mellitus. *Journal of Diabetes Science and Technology*, 3(2):269–282, 2009.

103. M. W. Percival, E. Dassau, H. Zisser, L. Jovanovic, and F. J. Doyle. Practical approach to design and implementation of a control algorithm in an artificial pancreatic beta cell. *Industrial and Engineering Chemistry Research*, 48:6059–6067, 2009.

104. Z. Ping-Jiang, E. D. Sontag, and Y. Wang. Input-to-state stability for discrete-time nonlinear systems. *Automatica*, 37:857–869, 1999.

105. L. S. Pontryagin, V. G. Boltyankii, R. V. Gamkrelizde, and E. F. Mischenko. *The Mathematical Theory of Optimal Processes*. Interscience Publishers, Inc.,

New York, USA, 1962.

106. V. M. Popov. *Hyperstability of Control Systems*. Springer-Verlag, Berlin, Germany, 1973.

107. A. S. Posnyak, E. N. Sanchez, and W. Yu. *Differential Neural Networks for Robust Nonlinear Control*. World Scientific, Danvers, MA, USA, 2000.

108. J. A. Primbs, V. Nevistic, and J. C. Doyle. Nonlinear optimal control: A control Lyapunov function and receding horizon perspective. *Asian Journal of Control*, 1:14–24, 1999.

109. G. Puscasu and B. Codres. Nonlinear system identification based on internal recurrent neural networks. *International Journal of Neural Systems*, 19(2):115–125, 2009.

110. G. Quiroz. *Bio-signaling Weighting Functions to Handle Glycaemia in Type 1 Diabetes Mellitus via Feedback Control System (in Spanish)*. PhD thesis, IPICYT (Potosinian Institute of Scientific and Technological Research), San Luis Potosi, Mexico, 2008.

111. G. Quiroz, C. P. Flores-Gutiérrez, and R. Femat. Suboptimal H_∞ hyperglycemia control on T1DM accounting biosignals of exercise and nocturnal hypoglycemia. *Optimal Control Applications and Methods*, 32:239–252, 2011.

112. B. Roszak and E. J. Davison. Tuning regulators for tracking siso positive linear systems. In *Proceedings of the European Control Conference*, 2007.

113. G. A. Rovithakis and M. A. Christodoulou. *Adaptive Control with Recurrent High-Order Neural Networks*. Springer-Verlag, Berlin, Germany, 2000.

114. E. Ruiz-Velázquez, R. Femat, and D. U. Campos-Delgado. Blood glucose control for type 1 diabetes mellitus: a tracking H_∞ problem. *Control Engineering Practice*, 12:1179–1195, 2004.

115. E. N. Sanchez and A. Y. Alanis. *Redes Neuronales: Conceptos Fundamentales y Aplicaciones a Control Automático*. Pearson Education, España, 2006.

116. E. N. Sanchez, A. Y. Alanis, and A. G. Loukianov. *Discrete-time High Order Neural Control*. Springer-Verlag, Berlin, Germany, 2008.

117. E. N. Sanchez, A. Y. Alanis, and J. J. Rico. Electric load demand prediction using neural networks trained by Kalman filtering. In *Proceedings of the IEEE International Joint Conference on Neural Networks*, Budapest, Hungray, 2004.

118. E. N. Sanchez and J. P. Perez. Input-to-state stability (iss) analysis for dynamic neural networks. *IEEE Transactions on Circuits and Systems—I: Fundamental Theory and Applications*, 46(11):1395–1398, 1999.

119. E. N. Sanchez, L. J. Ricalde, R. Langari, and D. Shahmirzadi. Rollover prediction and control in heavy vehicles via recurrent neural networks. *Intelligent Automation and Soft Computing*, 17(1):95–107, 2011.

120. P. O. M. Scokaert, J. B. Rawlings, and E. S. Meadows. Discrete-time stability with perturbations: application to model predictive control. *Automatica*, 33(3):463–470, 1997.

121. R. Sepulchre, M. Jankovic, and P. V. Kokotović. *Constructive Nonlinear Control*. Springer-Verlag, Berlin, Germany, 1997.

122. Y. Song and J. W. Grizzle. The extended Kalman filter as local asymptotic

observer for discrete-time nonlinear systems. *Journal of Mathematical Systems*, 5(1):59–78, 1995.

123. E. D. Sontag. On the input-to-state stability property. *Systems & Control Letters*, 24:351–359, 1995.

124. J. T. Sorensen. *A Physiologic Model of Glucose Metabolism in Man and Its Use to Design and Assess Improved Insulin Therapies for Diabetes*. PhD thesis, MIT, USA, 1985.

125. G. Sparacio, F. Zandariego, S. Corazza, A. Maran, A. Facchinetti, and C. Cobelli. Glucose concentration can be predicted ahead in time from continuous glucose monitoring sensor time-series. *IEEE Transactions on Biomedical Engineering*, 54(5):931–937, 2007.

126. G. M. Steil, A. E. Panteleon, and K. Rebrin. Closed-loop insulin delivery— the path to physiological glucose control. *Advanced Drug Delivery Reviews*, 56:125–144, 2004.

127. The Diabetes Control and Complications Trial Research Group. The effect of intensive treatment of diabetes on the development and progression of long-term complications in insulin-dependent diabetes mellitus. *New Englang Journal of Medicine*, 329:977–986, 1993.

128. Z. Trajanoski and P. Wach. Neural predictive controller for insulin delivery using the subcutaneous route. *IEEE Transactions on Biomedical Engineering*, 45(9):1122–1134, 1998.

129. A. J. Van Der Schaft. *L_2-Gain and Passivity Techniques in Nonlinear Control*. Springer-Verlag, Berlin, Germany, 1996.

130. T. Van-Herpe, M. Espinoza, B. Pluymers, I. Goethals, P. Wouters, G. Van den Berghe, and B. De Moor. An adaptive input-output modeling approach for predicting the glycemia of critically ill patients. *Physiological Measurement*, 27:1057–1069, 2006.

131. M. Vidyasagar. *Nonlinear Systems Analysis*. Prentice-Hall, Englewood Cliffs, NJ, USA, 2nd edition, 1993.

132. Y. Wang, E. Dassau, and F. J. Doyle. Closed-loop control of artificial pancreatic β-cell in type 1 diabetes mellitus using model predictive iterative learning control. *IEEE Transactions on Biomedical Engineering*, 57(2):211–219, 2007.

133. S. Wild, G. Roglic, A. Green, and R. Sicree. Global prevalence of diabetes. *Diabetes Care*, 27:1047–1053, 2004.

134. J. C. Willems. Dissipative dynamical systems: parts I and II. *Archive for Rational Mechanics and Analysis*, 45(5):321–351, 1972.

135. J. L. Willems and H. Van De Voorde. Inverse optimal control problem for linear discrete-time systems. *Electronics Letters*, 13:493, 1977.

136. J. L. Willems and H. Van De Voorde. The return difference for discrete-time optimal feedback systems. *Automatica*, 14:511–513, 1978.

137. R. J. Williams and D. Zipser. A learning algorithm for continually running fully recurrent neural networks. *Neural Computation*, 1:270–280, 1989.

138. D. Youla, L. Castriota, and H. Carlin. Bounded real scattering matrices and the foundations of linear passive network theory. *IRE Transactions on Circuit Theory*, 6(1):102–124, 1959.

Index

Note: Page numbers ending in "f" refer to figures. Page numbers ending in "t" refer to tables.